$6.41 Profit Per Hen Per Year
The Corning Egg Book

by Michael K. Boyer

with an introduction by Jackson Chambers

This work contains material that was originally published in 1910.

This publication is within the Public Domain.

This edition is reprinted for educational purposes
and in accordance with all applicable Federal Laws.

Introduction Copyright 2018 by Jackson Chambers

The World's Largest Selection of Vintage Poultry Books

www.VintagePoultry.com

Self Reliance Books

Get more historic titles on animal and stock breeding, gardening and old fashioned skills by visiting us at:

http://selfreliancebooks.blogspot.com/

Introduction

I am pleased to present yet another title on Poultry.

The work is in the Public Domain and is re-printed here in accordance with Federal Laws.

As with all reprinted books of this age that are intended to perfectly reproduce the original edition, considerable pains and effort had to be undertaken to correct fading and sometimes outright damage to existing proofs of this title. At times, this task is quite monumental, requiring an almost total "rebuilding" of some pages from digital proofs of multiple copies. Despite this, imperfections still sometimes exist in the final proof and may detract from the visual appearance of the text.

I hope you enjoy reading this book as much as I enjoyed making it available to readers again.

Jackson Chambers

WHERE THE $6.41 PROFIT PER HEN PER YEAR IS MADE

Interior Laying House No. 2. Fifteen hundred laying pullets in one house. The exterior of this building is shown on page 33. Note the raised canvas shutters fastened to the ceiling, allowing air and sunshine to stream in. No glass is used in this house. Since this photograph was taken the rafters have been ceiled with matched boards.

Contents

	PAGE
Introductory	5
Egg Farming as a Profitable Undertaking	9
Sunny Slope Farm	10
Over Six Dollars Per Hen Per Year	12
Premium Prices for Corning Method Eggs	13
Marketing the Eggs	17
The Breed to Keep	18
Pullets or Yearling Hens?	19
House Room and Size of Flocks	20
Buildings on Sunny Slope Farm	21
Fresh Air and Sunlight	22
Incubator Cellar	23
The Brooder House	25
Colony Houses	29
Main Laying Houses	31
The Breeding House	37
The Cockerel House	37
The Feedhouse and Workshop	39
Operating the Incubators	39
Ventilation and Moisture of Incubator Cellar	41
When to Hatch	43
Taking Care of the Chicks	44
Feeding Newly-Hatched Chicks	45
Feeding Pullets While on Range	47
Feeding Laying Pullets	48
Feeding Cockerels for Broilers	49
Feeding the Breeding Stock	51
Feeding Hens Through Molt	51
Mash—Morning or Night?	52
Fresh Cut Bone	52
Green Food	53
Drinking Water	55
Charcoal, Grit and Oyster Shell	56
Hard Coal Ashes	56
Eggs for Hatching	56
Cleanliness	57
Punctuality and Regularity	58
The Corning Method Applies to Small Plants	59

Introductory

When Edward and Gardner Corning, whose chronicles of success in egg raising are recorded in this little book, had decided that poultry offered the means of a good livelihood, they began to study all the literature they could find on the subject. They were soon impressed with the wonderful amount of common sense and wisdom contained in the writings of Professor Gowell, of the Maine Experiment Station. Edition after edition of his writings had been printed and almost as quickly exhausted. The Cornings secured all that he has written and studied it diligently and to a good purpose.

Their methods of work, their theories of profitable poultry raising are all here set forth, plainly and clearly. Their success will be an inspiration to every poultry raiser the country over. What they have done with a large flock can be accomplished with a few hens—the same principles apply.

When Prof. Gilbert M. Gowell's sudden death was announced (May 6, 1908), the poultry world sustained an irreparable loss. He gave more than twenty years of his life to the service of the University of Maine, during which time he won a reputation which made his work regarded by the United States Department of Agriculture as the most important ever carried on in this country in poultry experiments.

Professor Gowell taught the poultry world better egg production, and poultry husbandry has greatly profited by his unceasing and persistent efforts in that direction.

A year or two before he died, Professor Gowell paid the writer a visit, when he explained that he was about to resign his professorship of Animal Industry so that he might devote his entire energies to poultry investigations and breeding. At that time, he said, he was devoting half his time to conducting poultry investigations for the Maine Station, and the other half to his private business—the personal management of the Go-well Farm.

Being asked about his work at the latter place, he handed me a little booklet primarily used as an advertisement, but which contains much information concerning his efforts. Here are a few extracts:

"For more than twenty-five years I have bred Barred Plymouth Rocks for producing good brown eggs, by selections from the general stock. While that system of selection gave birds that laid eggs of good size, shape and color, there was no means of knowing whether the eggs incubated came from hens that were good or poor layers, and it was reasonable to suppose that as many chickens came from mothers who had laid poorly through the winter, as from those that had laid well. Indeed, recent investigations convince me that the eggs from hens that have only just gotten well under way laying at the commencement of the incubating season, yield more chicks than do those from hens that have been laying well since early fall.

Thoroughly believing in the principle of breeding performers to performers to get performers, I determined to rigidly cull out all non-performing hens, and breed only the good layers to the sons of good layers to get good layers. In order to do this, in 1898 I devised and constructed, at the Maine Experiment Station, fifty-two trap nests and commenced the selection of the best laying hens for foundation stock.

At the end of the year all birds that had not laid 160 eggs were rejected, and those that had laid above that number were retained for breeding. They were bred to sons of hens that had laid 200 or more eggs in a year.

This work of selection and breeding has been continued through every year down to the present time. Before the commencement of this attempt at improvement, the hens had averaged about 120 eggs each per year. The averages during the past two years have been above 144 per bird. I believe this increase in the average yields, of more than two dozen eggs per bird, is the direct result of the system of breeding practiced. The blood of the drone and average worker has been excluded for eight generations. The best have been bred to the best to get the best.

So thoroughly did I believe in this work of improvement in breeding and its possibilities, that I established Go-well Farm, that I might have a great breeding plant of my own, where I could carry forward the work in a larger way than I had previously been able to do.

Go-well Farm consists of about 100 acres, of which thirty acres are under cultivation or in grass or clover. Twenty-four incubators, with capacity of about 10,000 eggs, are used. There are forty brooder houses, each 7 x 12 feet in size, and high enough so a man can stand erect in them. Over 100 brooders are used. The house for laying birds is 20 x 400 feet in size, and required 100,000 feet of lumber and a ton and a half of nails in its construction. Something over 6,000 chickens are raised each year, and 2,000 pullets are kept as layers.

The leading purpose of this plant is the production of eggs for human food. They are expressed to market every day when perfectly fresh. Business economy demands that the food be changed over into as many pounds of chicken and as many dozens of eggs as is practically possible. It makes a difference in the balance sheet whether the hens have laid ten or twelve dozen eggs each in the year.

In order to be sure to breed the chickens from the best laying hens, I equipped the laying house with 400 trap nests, and they are used exclusively every day in the year.

The stock is grown in individual brooders, in the open fields, on clean grass land. In October they are put into laying quarters, in flocks of 100 each, with plenty of chance for exercise. They are never shut in closely.

During each of the last two years 6,000 chickens have been raised to maturity at Go-well Farm. About 12,000 eggs were incubated to produce them. The eggs were a little more than half hatchable during March and April. Those laid in May yielded more chicks. It must be borne in mind that these eggs were not from the general flock, but were from the hens that had been laying the heaviest during the winter. Tests show that the eggs from hens that have laid heavily during four or five months do not yield as many chicks as do those from hens that have laid but little.

During the first seven years' work in the development of this family, only those hens that laid from 160 eggs upward were used as breeders. For the last two years the general breeding has been done by selecting, in April, the pullets that had laid heavily since November. I made this change because it assured birds that had been doing their work in winter, when the higher prices for eggs prevail, and I wanted to fix the function of early as well as heavy laying.

It must not be forgotten that the fathers of every chicken grown, for eight generations, had mothers that laid from 200 to 255 eggs each in a year. There is not a chicken on Go-well Farm but has nine generations of this rigid selection of parents behind it."

Sunny Slope Farm has profited by Professor Gowell's advice. Although but three years in harness, the Messrs. Corning—father and son—have proved themselves apt scholars, and produced wonderful results, although in some ways they have departed from the Gowell Method. These departures we note throughout the book. The reader can make comparisons, and if he will carefully follow the teachings herein laid down, there is no reason why equal success should not crown his efforts.

MICHAEL K. BOYER.

PAPER BOXES USED BY SUNNY SLOPE FARM IN SHIPPING EGGS TO CUSTOMERS

$6.41 Per Hen Per Year

Egg Farming as a Profitable Undertaking

Egg farming is considered generally to be more profitable than any other branch of poultry-keeping, it being reasoned that the profits therefrom are surer and larger.

Success in any branch of poultry-keeping, whether for egg production, broilers for the table or exhibition specimens, requires infinite care, great regularity and close attention from daylight to dark, for seven days in the week and fifty-two weeks in the year.

The duties in any one branch are so multitudinous and exacting that the man or woman who can multiply them by three, or even two, and succeed on a commercial basis, is very rare.

In the limited number of instances where a dual plant has been operated at a profit, it has always been a question in the mind of the writer if the net returns would not have been much greater had only one branch been attempted.

Specialization is a characteristic of the age. It has found its way into every branch of human activity, with the result that the world is passing through the greatest period of its development, in every department. To succeed with poultry, specialization is necessary.

It is the intention of this book to describe one of the most successful egg plants in America, and to show how two novices in the short space of three years have built up a plant that is netting a profit of several thousand dollars a year—a net profit exceeding $6 per hen.

This is the plant at Sunny Slope Farm, at Bound Brook, N. J. It is owned by Messrs. Edward and Gardner Corning, father and son.

When they made up their minds to go into egg farming for a living, they read everything they could get hold of on the subject, particularly the valuable writings of the late Professor Gowell in connection with the Maine Experiment Station, adopted what appealed to their reason, rejected what did not measure up to standard by their rule, invented other plans and methods, and from the whole evolved the Corning Method of Poultry Keeping.

The plant and experiments have cost more than $20,000, but these men now have the satisfaction of knowing that what has been a dream to thousands is a reality to them, and that they have a business that will net them annual incomes on a par with other large and profitable enterprises, give them plenty of healthful exercise and absolute independence.

Any person who will closely follow the simple rules laid down in this book can make the same great success of poultry-keeping for eggs that Sunny Slope Farm has achieved.

As Sunny Slope Farm was built, partially, on a plan originated by the late Prof. Gilbert M. Gowell, M.S., in charge of poultry investigations, Maine Agricultural Experiment Station, it might be stated that for many years poultry work has been carried on at the University of Maine. It was not, however, until 1897 that the Station decided to begin a series of poultry investigations on a somewhat extended scale. Since 1904 this work has been carried on in co-operation with the Bureau of Animal Industry of the United States Department of Agriculture.—EDITOR.

Sunny Slope Farm

Sunny Slope Farm comprises twelve acres of land, the soil being of a reddish sand loam. From a line through the centre from east to west, the land slopes gently to both north and south. It is on this higher ground that the poultry buildings have been placed, and care was taken to have every one of them face due south.

The northern part, which lies alongside the turnpike, is used exclusively for raising wheat and clover for green food. During the summer season the southern part is devoted to the colony houses and range for the growing pullets.

A peculiarity of the farm is that it is not surrounded by a fence—there is not a fence on it except those yarding in the runs of the breeding and brooding houses respectively. Of course, an enclosing fence has no disadvantages, but in this instance it has no advantages, and besides it would have cost money which was more needed in other directions.

The first building as one enters from the road, is the office, feed room and chief attendant's home. This is a two-story building, described in detail in a later chapter.

South of this, seventy-five feet away, is the building used as a cockerel house, where the young males are housed as soon as they are old enough to determine their sex, and prepared for market.

To the east of the office building is the incubator cellar and brooding house. This is one of the most perfect houses of its kind in America, judging by the results obtained. Out of 3,313 chicks last season, less than ten per cent. was lost therein.

In line with the laying houses and to the west thereof, is the breeding house, capable of taking care of 350 breeding hens and their mates.

Twenty-five feet to the east of this is No. 1 laying house. This building is 100 feet long and 12 feet wide, but is to be enlarged this summer to 160 feet in length and 16 feet in width, so as to enable it to carry 1,500 layers.

To the south of No. 1 house is No. 2 house, where 1,500 layers through the winter months shell out the eggs in close to record quantities. After thoroughly testing it from every point, there is not one feature in its construction or arrangement that its proprietors would change. It is a perfect house for laying stock.

In front of all these buildings are the colony houses, in which the young pullets are matured after being removed from the brooder house. There are twenty of these, and they are moved as occasion demands.

No architect or master carpenter has had anything to do with the construction of these houses. The Messrs. Corning did all the work of planning the buildings, and with the aid of unskilled helpers erected them.

There is a splendid driven well of sparkling water on the farm, 117 feet deep, and, by the aid of a windmill and tank, it is piped to all necessary parts of the plant.

There is not a tree on the place and the shade for the growing stock has to be artificially furnished.

Sunny Slope Farm is not a believer in fences. The late Professor Gowell was, according to the Maine Station Bulletins, and, without wishing to criticise the Sunny Slope Farm's ideas in the matter, years of experience in breeding show that fowls allowed out in the open, on terra firma, will be more rugged, have a brighter look, and suffer less from colds than hens continually housed. The State of Maine is subject to long spells of severe cold weather, with the temperature considerably below zero at night, and above zero during the day, and with a good deal of high wind. To these laying houses at the Maine Station, the yards are only on the north side, as the birds are kept in the building until the weather is suitable for opening the small doors in the rear wall. The necessity of getting the fowls out of the open-front house, where they are really subject to most of the out-of-door conditions during the daytime, is not so great as when confined in closed houses with walls and glass windows. There is no glass used in the laying houses on Sunny Slope Farm.—EDITOR.

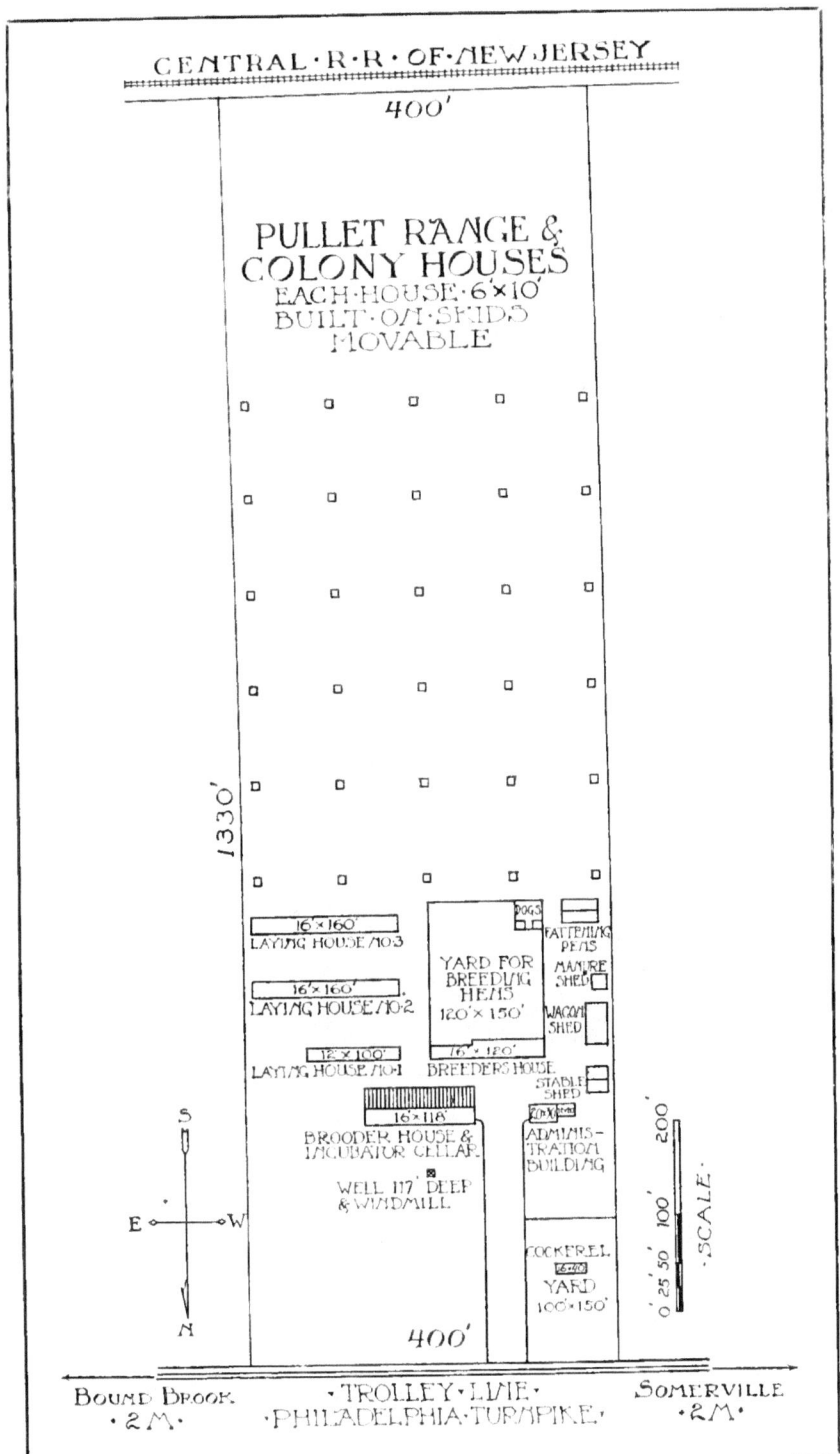

Plan of Sunny Slope Farm

Over Six Dollars Per Hen Per Year

The layers on Sunny Slope Farm are making a net profit of over $6 each for their owners. This is over the cost of incubating, feeding, marketing and hired help.

No exceptional methods are employed in making this profit. This farm was established to produce eggs for table purposes, and this is the main source of profit. The surplus cockerels and the pullets after completing their laying season are simply treated as by-products, and are disposed of as quickly as possible.

Every effort is bent toward a large production of eggs, especially in the winter months, and it is the success which has attended these methods which has brought the big margin of profit named.

Accounts are carefully kept on Sunny Slope Farm, and the profit or loss on any day's business can be readily told. These books are kept in the same methodical way that everything on the farm is looked after.

The exact amount of food consumed each day is kept on record. For instance, on January 12, 1908, the feed used by 1,953 pullets in the laying houses and 210 breeders was as follows, with the cost of same:

150 lbs. cracked corn and wheat		$2.40
52 lbs. oats		.85
26 lbs. meals	$0.44	
13 lbs. ground oats	.20	
26 lbs. bran	.37	
150 lbs. cut bone	1.50	
Gasoline for engine	.08	
		2.65
35 lbs. cut clover		.35
Grit, shell and charcoal		.15
		$6.40

This was an average day, and shows that the cost of feeding each hen runs a little less than .03 of a cent per day in the winter.

In the summer and fall months the cost for clover or green food is practically wiped out, as much other food is not needed to supply the fowl's requirements. This reduces the cost to very little above a quarter cent per hen per day for feed.

To keep a pullet for ten months after reaching the laying point at this rate will cost 86 cents for feed. To this should be added the cost of hired labor, which brings the total cost of keeping a hen on this farm through her first laying period, $1.11.

It requires something less than 40 cents to raise a pullet to the laying period, including cost of incubation and hired labor.

At the present time, summer of 1909, the constant services of two men, in addition to those of Edward and Gardner Corning, are required to do the work, and also the services of a boy—one-half of each day—to assist in gathering and packing eggs.

This makes the cost of keeping a pullet up to the point where she has completed her first laying season just $1.50.

It costs approximately 15 cents to raise a Leghorn cockerel to the broiler size, when they are worth about 30 cents each, alive.

The secret in securing the advanced prices is the fact that the Sunny Slope Farm caters to practically a retail or select trade. The low cost they record for feeding stock, or raising pullets to ten months of age, or getting cockerels up to broiler weight, can only be obtained by purchasing feed in large quantities.

The smallest details are attended to with fixed regularity, each employee on the farm strictly attending to his work.—EDITOR.

Once a strain of birds has gained a reputation for heavy egg production all the stock raised can be sold at remunerative prices. This is particularly true of the females, for which there is a ready market at $2 each. All the females sold on this farm have been disposed of at this figure.

The pullets last season averaged 143.25 eggs each for the ten months from December 1st to September 30th, and are doing even better this season.

These eggs were disposed of at prices as high as 65 cents per dozen, and never for less than 40 cents, averaging nearly 50 cents a dozen. Contracts for the entire yield of eggs have been made guaranteeing these prices for the next year.

The product of 1,953 pullets was 279,792 eggs, or 23,316 dozen.

This gives the following result:

REVENUE

23,316 dozen eggs at 49c. (average price)	$11,424.84
1,900 pullets as breeders, at $2	3,800.00
800 live broilers at 30c	240.00
Manure	250.00

$15,714.84

EXPENDITURE

Raising 1,953 pullets to laying point	$781.20
Maintaining 1,953 pullets through laying season of 10 months	2,167.83
Raising 800 cockerels to broiler size of 1½ lbs	120.00
Cartons, postage, etc.	125.00

3,194.03

$12,520.81

Leaving a net profit of $6.41 per head of laying stock.

Premium Prices for Corning Method Eggs

There is always a demand at a big premium over market prices for eggs that can be depended upon—eggs that are known to be fresh and of uniform good quality.

In the bigger cities this demand is more pronounced than in smaller places, but the demand for fancy first-quality eggs is universal. In the big cities a premium ranging as high as 40 to 50 cents a dozen is paid for eggs of this quality, doubling the retail price of them in the late fall and winter months.

Except in the summer months it is very difficult to get eggs that can be depended on for table use. The demand is so great, and the price paid so profitable, that many poultrymen cannot resist the temptation to mix their first-quality eggs with cheaper ones.

Several dealers in the East make a practice of gathering the eggs in the cold months from the farmers weekly and disposing of them to the commission men. These eggs are two or three weeks old, as a rule, before they reach the consumer, by which time they are no longer fresh enough for table use, and particularly so because they have almost invariably been fertilized. Yet they are everywhere sold as "near-by fresh" eggs, and are graded ahead of "western fresh," although in reality they are seldom any better.

A New York matron, shortly before Christmas, bought a quarter's worth of these eggs at the rate of 60 cents a dozen, for the purpose of making a cake for a special occasion. The batter was prepared and at the proper time the eggs, one by one, were broken in; the first four seemed all right, but the fifth was bad, and the whole lot, eggs, batter and all, had to be thrown away.

A grocer in a Jersey city had a call for newly laid eggs for a sick patient, regardless of cost, the only stipulation being that they must be absolutely fresh. An hour or two later a farmer walked into his store with three dozen eggs which he said had been laid the day before. He was paid 50 cents a dozen for them and they were sent by special messenger to the sickroom, the grocer congratulating himself that he had been able to accommodate his customer so promptly. It turned out, however, that every one of these eggs was rotten. On investigation it was found that the farmer had gathered them the day before he brought them in, but they were apparently a lot that hens had been sitting on

The quotations on eggs in the New York market of January 5, 1910, are here shown. It should be noted that the best prices obtained are nearly three times the poorest price of 18 cents per dozen. About this time the Cornings were getting 75 cents a dozen from fancy retail trade in New York City.

the previous fall but that had failed to hatch and had not been discovered for weeks after.

It is circumstances such as these that make the housewives of the East willing to pay a big premium for dependable eggs.

Eggs kept in cold storage rapidly lose their flavor. If they are thus kept from spring to winter they have a taste akin to a mixture of cornstarch and water, often with a little musty flavor added.

To show the relative prices, January 5, 1910, in the Philadelphia market, eggs were selling in one of the big grocery stores of Philadelphia as follows:

Non-fertile White Leghorn eggs.............................60 cts. per doz.
"Milhen" eggs in paper boxes................................50 cts. per doz.
Nearby fancy brown and white eggs in bulk..............50 cts. per doz.
Storage eggs..35 cts. per doz.

The Leghorn eggs were advertised "for invalids." The "Milhen" eggs are being extensively advertised in the Philadelphia newspapers. They are packed in attractive cartons of one dozen each, sealed and dated. The box purchased contained mixed eggs as to color.—EDITOR.

Every one is suspicious of an egg that is supplied on hotel or restaurant tables. Yet people are willing to pay good prices for this delicacy if its freshness and quality can be guaranteed. The hotel men know this, and eggs that can be depended on have an unlimited market at premium prices amongst the first-class hotels and restaurants.

The steward of a large New York hotel, in common with those at the head of other high-class restaurants, has had great difficulty in obtaining fine-quality eggs for his tables.

"We don't worry any more," he said the other day. "We are now getting dependable eggs, and we *know* that every egg that goes on our tables is fresh and sweet."

His eggs are now coming from the Sunny Slope Farm, produced by the Corning Method. They are costing him 10 cents a dozen more than the highest market price for "near-by" fresh eggs, but he is well content to pay it.

One of the stockholders of a prominent hotel in New York ordered eggs for breakfast there, and after he had eaten them he sent for the steward.

"Where did you get those eggs?" he asked. "They are the finest that I have ever eaten."

"We are not telling where we get them, but we can supply you with them regularly," was the reply.

The steward had just arranged for a daily supply of eggs from Sunny Slope Farm.

Contracts have been made with hotels for a stipulated number daily, the year through, at a premium of 10 cents a dozen over the highest quoted price for strictly "near-by" new-laid eggs, and a guarantee that the price shall not go below 40 cents a dozen during the year. On this basis as high as 65 cents a dozen was realized in November and December, and the price averaged about 50 cents for the twelve months.

Even with these terms Sunny Slope Farm cannot begin to keep pace with the demand. The largest grocers in New York City, who supply the most exclusive trade, recently made an offer for the entire output of Sunny Slope Farm eggs on the terms above stated, but previous contracts prevented the acceptance of this offer. In the spring of the year, when hens are laying at their best, the commission houses pay a substantial premium for all surplus eggs from this farm, in spite of the fact that the market is at this time fully supplied with so-called fresh eggs.

The demand for dependable eggs from stock wholesomely fed is unlimited among the restaurants in New York and other large cities. There is no likelihood of the supply equaling the demand in the next decade, the increase in population in the cities more than keeping pace with the increased egg yield.

A much higher price can be obtained by supplying the residential trade privately, but the expense of marketing is considerably greater. It has the further disadvantage of falling off almost entirely in the summer months when the richer city residents close their houses and go to their country places or the watering resorts.

The restaurants in New York which are supplied with Corning Method eggs are famous for their egg dishes. Their sweet, wholesome, palatable taste is at once remarked. A better price can be and is obtained for dishes prepared with them than these same restaurants could previously get.

Sunny Slope eggs, of course, are white in color, and average 25 to 26 ounces to the dozen.

In order to guarantee strictly fresh eggs, the Sunny Slope Farm has the eggs collected four times a day. Shipments are made daily. To secure such fancy prices, however, it must be understood that good business methods had to be employed. Mr. Corning, Jr., secured the trade. He proved the quality of his goods. The rest was easy.

OFF TO MARKET

Showing one method of packing Sunny Slope eggs

Marketing the Eggs

Common sense combined with good business methods must be used in marketing the eggs, if it is desired to obtain the high prices referred to in this book. At Sunny Slope Farm the minutest detail is looked after to have the eggs reach the market not only with the best possible appearance, but of the highest quality.

They are packed a dozen in a box, and care is taken that the eggs in each box be of uniform size. This does not take much time and it counts for a great deal in the looks of the eggs.

As each basket of eggs is brought from the laying houses they are carefully gone over and every particle of dirt is removed. This adds much to the appearance of the eggs, and is necessary to prevent the contents of the shell from being tainted. An egg-shell is very porous and the flavor of the meat therein is readily affected by contamination.

Up to the present Messrs. Corning have been able to dispose of their entire supply to the big hotels, and they have had to use no care to see that their customers really get their eggs, for the reason that they are actually delivered to them by their own employees. Were they developing a retail trade or supplying a market at a distance they would use seals on every box, and on each would be printed clearly and plainly the fact that the egg was produced by the Corning Method, and on it would also be stamped the day on which the eggs were laid. In this way a consumer would be guaranteed definitely that he was eating a fresh egg.

The seal is necessary to protect the producer from unscrupulous dealers who might refill the boxes with inferior eggs and sell them as Corning Method eggs, when they reach the consumer through the retailer.

Trade is attracted, first, by appearance; second, by quality. Care is taken to sort the eggs according to size, and to make their appearance tempting to the appetite. This gave a reputation that made possible a demand greater than the supply.

EGG RECORD—JANUARY, 1908, TO JULY 1, 1909

1908.	Dozens of eggs.	Average monthly prices	Total.
January	2,386	$0.55	$1,312.30
February	2,440	.53	1,293.20
March	4,050	.52	2,106.00
April	3,165	.50	1,582.50
May	3,574	.45	1,608.30
June	3,150	.40	1,260.00
July	2,071	.42	869.82
August	1,795	.45	807.75
September	1,176	.50	588.00
October	787	.55	432.85
November	1,253	.58	726.74
December	2,044	.60	1,226.40
1909.			
January	2,743	.61	1,673.23
February	2,806	.59	1,655.54
March	4,657	.49	2,281.93
April	3,639	.45	1,637.55
May	4,109	.44	1,807.96
June	3,622	.41	1,485.02

January 1 to September 15, 1908, an average of 1,953 layers.
September 16, 1908, to July 1, 1909, an average of 2,215 layers.

September 1, 1908, transporting pullets from the range into the laying houses was begun.

From September 1, 1908, yearlings were sold as breeders, except those selected for the breeding pen.

The Breed to Keep

On Sunny Slope Farm only Single-Comb White Leghorns are kept. There was no sentiment in the selection of this breed. It was settled that the farm was to make a specialty of eggs for table use, and for this purpose the weight of opinion and experience seemed to point to the White Leghorns.

There is a difference in the preference of the respective markets as to the color of the shells—New York prefers a white shell, Boston a brown shell. This preference should be given consideration in the selection of a breed.

Professor Gowell gave the preference to the Barred Plymouth Rocks at the Maine Station, and obtained excellent egg results. The average of eight years from the Barred Rocks was 134.27 eggs for twelve months at the Maine Station, while the Leghorns on Sunny Slope Farm last year averaged 143.25 for ten months, at the rate of 171.9 a year.

Kept on the same principle as that employed at Sunny Slope Farm, these Barred Rocks would probably have done better than they did in the Maine Station. Some of the details might have to be changed. Each 20-foot roosting closet will hold only about 175 Rocks as against 200 Leghorns. The amount of animal food might have to be reduced to prevent the Rocks going too much to flesh.

With such simple modifications to meet the characteristics of the breed chosen, any of the Plymouth Rock, Wyandotte or Brahma fowls, as well as other families of the Mediterranean breeds, may be profitably kept as egg producers.

The Rocks will lay eggs somewhat larger, it is claimed, than the Leghorns (although this is doubtful) and their cockerels, as well as the layers who have passed the profit line, will bring larger returns for meat, if they are disposed of for this purpose.

On no consideration should half-breeds or mongrels be kept. There is no money in them.

The strain is of much more importance than the breed. A hen lacking in physical vigor will not make a good egg machine. There are some lines of White Leghorns which have been so inbred that they do not lay a large number of eggs, and will not produce strong chicks nor give any large percentage of fertility.

Great caution was exercised in this regard by the owners of Sunny Slope Farm. Their birds are uniformly large in size, have great vigor and are believed to be the greatest egg-producing strain in the world.

It is most essential, for success, to hand down the qualities of one strain of birds, which have been produced. If, however, the old theory of the introduction of new blood is accomplished by going outside for males, so as to avoid inbreeding, all possibility of handing down the virtues of mother to daughter, and of father to son, is lost, and time is wasted. In order to continue all the qualities which have been devel-

---

Sunny Slope Farm having New York City as its market, where the best prices are realized for white-shelled eggs, no doubt wisely chose the White Leghorn fowl.

Professor Gowell, in a report, says: "At the time we began this work we were carrying three breeds: Barred Plymouth Rocks, White Wyandottes and Light Brahmas. With the particular strains that we had of these breeds, the Barred Plymouth Rocks seemed the most promising, and the work here reported is with this breed. As the New England market demands large, dark-brown eggs, only birds laying such eggs have been used for breeding."—EDITOR.

oped in a given strain, line breeding must be strictly adhered to, and yet inbreeding must be avoided. Sunny Slope Farm's method is this:

A small pen is made up of selected yearling hens, and one cockerel for each 12 hens is added to this pen.

The eggs gathered from this pen are set in incubators containing eggs from it only.

Before placing the chicks, resulting from such hatching, in the brooder house, a hole is punched in the web of one foot of each chick. When the cockerels are collected from the range, those found with a hole in the web of one foot are set apart and grown to maturity—the choicest ones being selected to head the next season's breeding pen. The pullets with a hole in the web of one foot go into the laying houses with the other pullets, and are disposed of as yearling hens.

When a sufficient number of chicks have been hatched to provide about 200 selected cockerels, needed for the next year's breeding, this pen is broken up—the hens going to a laying house—where they produce sterile eggs for market, while the cockerels which were mated with them go to the sales department. By following this method, line breeding is adhered to and inbreeding avoided.

Pullets or Yearling Hens?

Only pullets are kept in the laying houses on this farm. When a fowl completes her first laying season, which, with a Leghorn, covers about ten months from the time she starts, she is sold off the place, unless her qualities are such that she is required in the breeding house. In this event she is kept a year longer.

A pullet will invariably lay more eggs than a yearling hen, and she will lay them in the winter months, when they will bring the highest prices. This has proved true every season on Sunny Slope Farm.

The first six months of last year the yearlings in the breeding house averaged 69 eggs each, and more than half of these were laid in April and May. They laid practically none in November or December, not being through their molt, and fell right off in July.

In the same period the pullets shelled out the eggs at the rate of 105 each, their heaviest month being March. They were laying well in November and continued without a break until the following September.

Both the yearlings and pullets were given the same rations and the same general treatment and care, except that the yearling flock was smaller and for three months the males ran with it.

Young hens sell readily as breeders and will bring a better price than a two-year-old hen. This is another reason for disposing of the pullets at the end of their first laying season.

In Bulletin No. 130, issued (June, 1906,) by the Maine Station, Professor Gowell says: "For the last seven years we have gotten the first eggs when the pullets were from four months and ten days old to four months and twenty days old. There is some danger of the pullets getting developed too early, and commencing laying too soon for best results, under this system of feeding. In order to prevent such conditions, the houses should not be located too close to each other, or to the feed troughs, and a large range should be given them so they may be induced to work, which they will do if given the opportunity early after their removal to the fields. Should the birds show too great precocity, and that they are liable to commence laying in August, the supply of cracked corn in the feeding trough is reduced, or taken away altogether, which causes them to eat the wheat, oats and dry meal instead, and they continue to grow and develop without getting too fat and ripe."

House Room and Size of Flocks

Economy of space and labor is one of the great questions of profitable poultry-keeping.

The larger the flock the greater the economy. This has been recognized from the first, and it is for this reason that the typical farmer has always kept his poultry in one flock. The old-time farmer, however, did not understand the first principles of profitable poultry-keeping, and consequently his chickens were never considered, and probably were not, a profitable asset of the farm. It was known that people living in cities and towns with smaller flocks generally secured a much larger egg yield per bird. This probably led to the growth of the opinion that hens should be kept in small flocks.

The late Professor Gowell, of the Maine Agricultural Experiment Station, was one of the first to realize that fowls could be much more profitably kept in larger flocks. He experimented along this line for several years and the results of these experiments have been considered so valuable that the United States Government has had them distributed freely throughout the entire country in the form of special Bulletins.

Prof. W. R. Graham, of the Guelph, Ontario, Agricultural Farm, and recognized as one of Canada's most expert poultrymen, said to the writer one day:

"I am keeping as many as 300 hens in one flock and am having splendid success with them. I believe that the large flock is the best for profitable poultry-keeping.

I am afraid, however, to recommend to the Canadian farmers and poultrymen that they keep their poultry in large flocks for several reasons. One is, the danger that is liable from disease, unless the most constant care is given to the birds. Another is, the fact that so few poultrymen or farmers will really give their birds the constant care that is needed. Without care, no plant can be made the highest success. Given the care necessary, the larger the flock the more profitable the plant will likely be."

On Sunny Slope Farm the large flock idea is developed from the moment the chicks leave the incubator. One hundred chicks approximately are placed in each hover. Three weeks later they are moved to the nursery pens, which are also located in the brooder house, the chicks in two hovers being combined and placed in one nursery yard. Thus the flock has grown to two hundred.

When they leave the nursery for the colony house the birds are old enough for their sex to be distinguished and the cockerels are sepa-

---

Professor Gowell, in "Poultry Investigations at the Maine Agricultural Experiment Station," published by the United States Department of Agriculture, says: "The cost of housing poultry is a very important item to the poultryman, and the amount of floor space required by each hen is a question much discussed and worthy of the most careful consideration and investigation.

Each of the pens in house No. 1 has 160 feet of floor space. When occupied by twenty-two birds, each individual has a floor space of 7.3 square feet. Each of the seven pens in house No. 2 has 240 square feet of floor space, giving each of the fifty pullets 4.8 square feet. In house No. 3 the four pens are twice as large as those of house No. 2, each containing 480 square feet. In each of two of the pens 100 pullets are kept, having 4.8 square feet of floor space to a bird—just the same allotment that is given in the pens of fifty birds in the No. 2 house. The 150 birds kept in each of the two other pens have only 3.2 square feet to a bird.

So far as health and egg production are concerned, thus far there is little to choose between the pens containing 2,250 and 100 birds, with 7.3 and 4.8 square feet to a hen. The fowls in the 150-bird pens, for reasons which are not attributed to the increased numbers or diminished floor space, did not do so well in 1904-5 as those in the other pens."

In 1907 Bulletin (No. 144) Professor Gowell reports "all the hens in the open-front houses, in flocks of 50 or 100, averaged 144 eggs each last year, and the birds were in excellent health. The front curtains were open all of the time every day, except the stormiest in winter." Again, later: "The health of the birds in a flock of 150, in comparison with those in the flock of 100 in like-sized pens, was apparently as good. In the pens of 50, 100 and 150 birds the proportional losses did not materially differ, being very small in all pens."—EDITOR.

rated and sent to the fattening pens or to be matured as breeders. This usually leaves from 200 to 250 pullets in each two nursery pens, and these are combined and put in one colony house. The colony houses as used on Sunny Slope Farm are large enough to easily accommodate this number of birds. Going through at night with a lantern there was always plenty of space, even with so many as 250 birds in a single house. Young pullets lie very close together, even in warm weather.

When the pullets are moved to the laying houses, the number is limited only by the capacity of the house, and here again the managers of this farm have economized in the matter of space to a point which would have been considered suicidal ten years ago. In No. 2 laying house, which is 160 feet by 16 feet, there are 1,500 laying pullets. Leaving the dropping boards out of consideration, this gives each bird a floor space of about 1.7 square feet; but with the dropping boards placed three feet from the floor, giving the fowls free access thereunder, they should be considered in the floor space, which brings the amount for each bird up to about 2.33 square feet.

This amount of floor space per bird would be out of the question if the house was divided into pens of the old-time regulation size and the number of birds equally apportioned thereto. One hundred and eighty-eight hens in a pen 20 feet by 16 feet would give the same floor space to each as to those in the larger house on Sunny Slope Farm, but the birds would be a great deal more confined. Each bird would have only 35 square yards to roam over, while those in the Sunny Slope laying house are accorded 280 square yards. It is true that the latter has to share this space with 1,499 neighbors which she meets in her wanderings from one end of the building to the other, while the former would only have to divide up with 187 neighbors. This would not add to her comfort or contentment, but the closer confinement would, if anything, reduce them. A hen seems to enjoy the greater liberty.

A few years ago it was considered that a fowl should have at least 10 square feet of houseroom floor space, and Professor Gowell in his most radical experiments never considered less than 4.8 feet. This, of course, was on the basis of a much smaller laying room.

In No. 1 laying house on this farm, which is 100 feet long by 12 feet wide, there are 450 laying hens which have a floor space more than twice that of the pullets in No. 2 house, or nearly 5 feet. The egg records show that there is practically no difference in the yield per hen in either house, but if there is any difference it is in favor of No. 2 house.

This economy in space cuts the cost of housing in two, which is a very important consideration as it lessens the amount of capital required by just that amount.

There is no doubt the size of the flocks could be increased much beyond the 1,500 point, but there would be little advantage. The housing cost would not be lessened to speak of, and it is questionable whether anything in time or labor would be saved.

The number of layers on Sunny Slope Farm is to be increased the coming season from 2,000 to 4,000, and the extra housing accommodation will be provided by erecting new houses on the same plan as the present No. 2 house, with a capacity of 1,500 each.

Buildings on Sunny Slope Farm

The buildings on Sunny Slope Farm are put up on a very substantial plan and on a principle that is original and unique.

At the same time there is nothing about them that could be dis-

---

Since the above account was written additions have been made to the buildings at Sunny Slope Farm, as follows:

Addition to brooder house 16 x 68 feet, giving accommodation to 7,000 chicks

pensed with. Economy was kept constantly in mind in their planning and erection, but was never allowed to interfere with effectiveness and simplicity.

Much of the wonderful success of this farm as a money-maker is due to the principle on which these houses are built.

The nucleus of the principle was discovered by the late Professor Gowell, but it is said he was about to abandon one of the essential features of the principle as unworkable, when his end came. That was the idea of having 100 lineal feet or more of laying house in one room. The Professor experimented for years with it, reluctant to give it up because he believed he was on the right track, but was never able to abolish draughts entirely therefrom and his birds were always troubled more or less with colds. In despair, he advocated a close partition house in his latest writings, making the rooms or pens not more than 20 feet long.

The objections from draughts in the Gowell houses have been met and entirely overcome in the Sunny Slope houses, without abandoning the important advantages of the single-room laying house.

Another change in the construction of the Sunny Slope houses compared with those on the Gowell Farm is that of placing them on posts five feet off the ground. This eliminates every possible danger of dampness, and with 2,000 layers this winter, notwithstanding the changeableness of the weather, not a single fowl has been found suffering even from a cold.

Fresh Air and Sunlight

Plenty of ventilation is essential in every house used on a poultry farm, from the incubator cellar to the laying houses.

When the supply of fresh air in the hen house is limited, when the oxygen that the fowls breathe is measured, their vitality is never robust and they die prematurely. Constitutional vigor is a matter of fresh air, as well as exercise, food and inherited stamina. Heavy egg production

---

at one time, and providing a cellar where sprouted oats are grown in winter for green food.

Addition to breeding house 16 x 72 feet, enabling them to carry 900 breeding hens and mates.

Additional laying house 16 x 160 feet, capable of carrying 1,600 layers. This is a reproduction of the old laying house and is believed to be the most perfect of all known houses for this purpose. The three laying houses are shown on page 33.

Farmers' Bulletin No. 357, entitled "Methods of Poultry Management at the Maine Agricultural Experiment Station," says: "During the summer of 1905 the management of a commercial poultry plant in Orono built a curtain-front house to accommodate 2,000 laying hens. This was built in accordance with unpublished plans prepared by the Maine Experiment Station. It represents the latest development of this style of house. The house is 20 feet wide by 400 feet long, and is divided into twenty sections, each being 20 feet square.

At first it was thought the house should be narrow, so it might dry out readily, but the 20-foot house dries out satisfactorily, as the opening in the front is placed high up, so that the sun shines in on the floor to the back in the shortest winter days.

The economy in the cost of the wide house over the narrow ones, when space is considered, is evident. The front and back walls in the narrow house cost about as much per lineal foot as those in the wide house, and the greatly increased floor space is secured by building in a strip of floor and roof running lengthwise of the building. The carrying capacity of a house 20 feet wide is 66 per cent. greater than that of a house 12 feet wide, and it is secured by merely building additional floor and roof. The walls, doors and windows remain the same as in the narrow house, except that the front wall is made a little higher."—EDITOR.

Although Professor Gowell did not invent the curtained-front idea for a house, he deservedly won the credit of greatly improving upon the principle. Since the introduction of curtain-fronts, the question of how to ventilate is seldom asked. There is no better method of ventilating, nor more sane way of disinfecting a house than by such means as will constantly admit the fresh air—through the curtains when closed, and through the open space when the curtains are hung up—and at the same time allow the sun's rays to seek every part of the interior.—EDITOR.

is principally a problem of how to maintain laying hens in a high state of health.

The oxygen in fresh air helps to maintain robust constitutions in the fowls; it gives red combs and bright eyes, it means little susceptibility to disease or colds, and it aids in the assimilation of the food.

A continual supply of fresh air results in a continual evaporation of the dampness in the house. The air exhaled from the fowl is laden with moisture from the throat and lungs. In cold weather this condenses in an airtight house or one built on ordinary lines and makes the whole house damp. The unhealthfulness of a damp house is unquestioned.

Another necessity in a hen house is sunlight—direct rays of the sun. Without sunlight no living organism can thrive, and this is true of the hen. On Sunny Slope Farm close attention has been paid to these factors.

The laying and breeding houses are all built on the curtained-front principle. In each 20 lineal feet of house there is a curtained window 3 feet by 9 feet. Medium-heavy cotton duck is used in these windows, and it is kept dusted, so that the air can continually circulate through it. In the daytime, except when it is storming from the south, these windows are opened and the sun can always find its way into every foot of the house.

Owing to the dryness of the atmosphere, not a single comb has even been nipped in these houses, and roup has yet to make its entry.

Sunlight and fresh air are furnished in abundance, all draughts are excluded and the secret of correct poultry housing is the result.

In another chapter the method of ventilating the incubator cellar and brooding house and admitting the sun to the latter is described. There is nothing so good for little chicks as warm fresh air and sunshine.

The temperature does not appear to affect the egg yield in these houses. The pullets get their blood warmed early in the morning, and, whether it is mild or zero weather outside, they continue to shell out the eggs.

Incubator Cellar

The incubation and brooding houses on Sunny Slope Farm are built in combination, and in outward appearance closely resemble similar houses on many of the more successful poultry farms in this country. Its interior arrangement, however, is entirely original. The incubator cellar is 50 feet long and 16 feet wide, with walls and floors of solid concrete 8 inches thick and with 8 feet between the floor and the ceiling. It is very important, in order to secure proper ventilation, to have the ceiling high enough.

The floor is 4 feet below the ground level, and in selecting a site for this house care was taken to pick the dampest spot on the farm. In close proximity thereto is a magnificent spring of the purest water, and conduits from this spring have been constructed underneath the incubator cellar in order to insure the necessary dampness. Water runs

---

On Sunny Slope Farm, in selecting a site for the incubator cellar, Mr. Corning told me "care was taken to pick out the dampest spot on the farm." The soil of Sunny Slope Farm is a very warm, dry gravel and sand loam. Professor Gowell, in Bulletin No. 90, says "a damp cellar would be poorly adapted for incubators." In 1905 an incubator cellar was constructed at the Maine Station measuring 30 feet square, 7 feet high in the clear, 5 feet of which is below the level of the outside ground. It is lighted by six three-light windows carrying glass 10 x 16 inches. The cement walls are finished smooth, and the cement floor is slightly inclined toward the southeast corner, where the intake of the drain is located, enabling free use of water from hose in cleaning the room. Two chimneys extend to the basement floor, containing ventilating flues. The room contains eighteen 360-egg machines, and by a little crowding would hold twenty-one.—EDITOR.

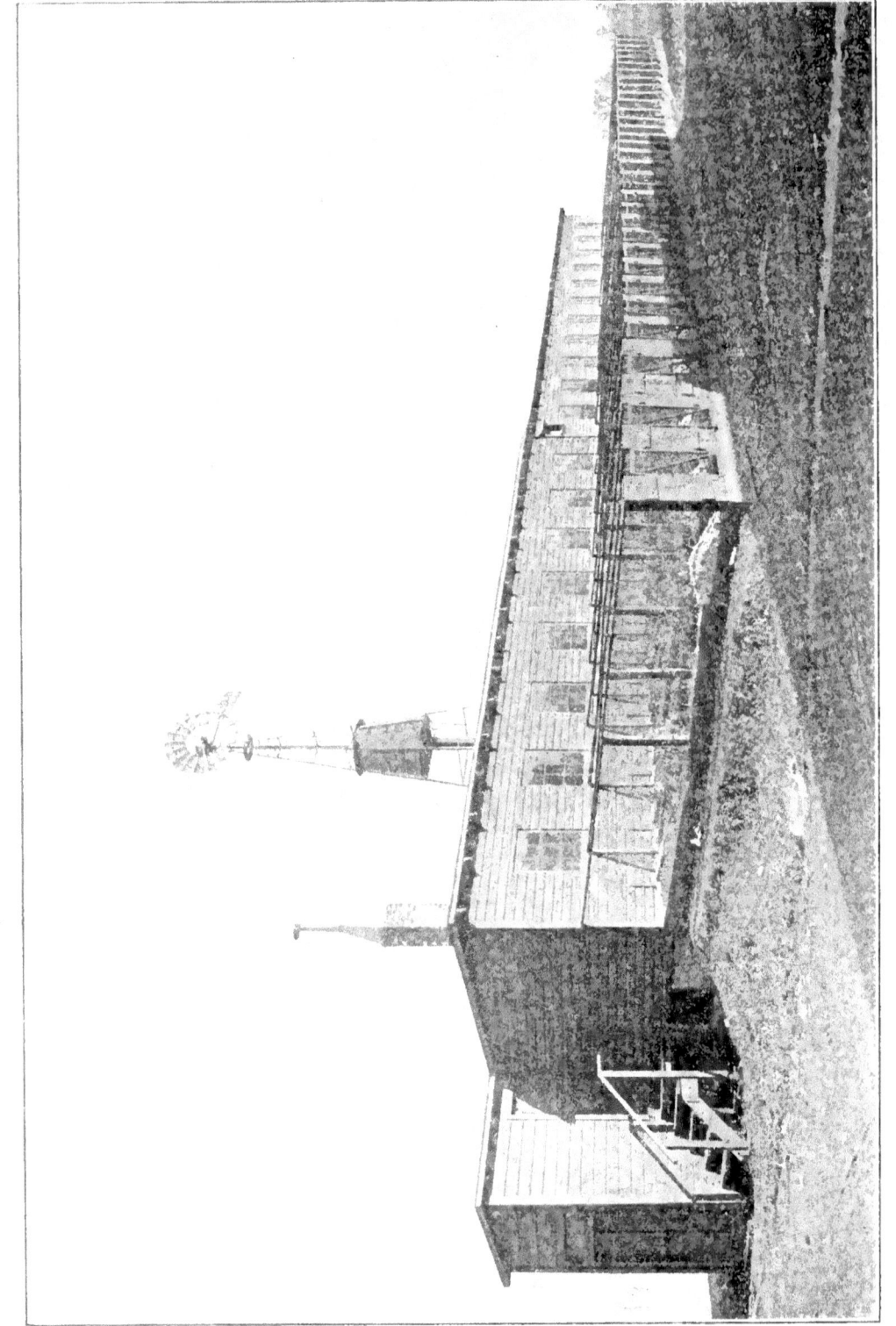

THE BROODER HOUSE

Entrance to the Incubator Cellar is shown on level with ground. Sprouted Oats Cellar is in far end of the building

constantly throughout the hatching season under the concrete floor of this cellar.

Entrance to the brooder house is obtained through a door in the vestibule in the west end of the building. From this vestibule are stairways leading down to the cellar and up to the brooder house.

Immediately inside of the vestibule is a fireproof chamber for the hot-water heater and adjoining is a large bin for coal.

Care has been taken to guard the incubator cellar against any extreme

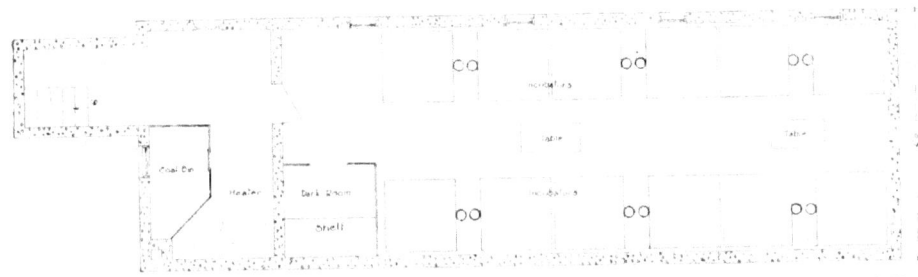

Plan of Incubator Cellar. The exterior of this is shown on the opposite page.

drops in temperature during the period that incubation is in progress. For this purpose two 2-inch emergency pipes from the heater run the whole length of the cellar to guard against freezing temperature therein. As a matter of fact, these emergency pipes have never been required, but the proprietors feel a greater degree of safety than they would if they were not there.

In one corner of the cellar is a dark room for testing the eggs. Therein is a counter big enough to take two trays set side by side. As the eggs of the one are tested and found fertile they are placed on the second tray. If infertile they are placed in a basket on the floor.

The Brooder House

The brooder house is 118 feet long and 16 feet wide, extending 68 feet beyond the incubator cellar. The first 50 feet, or that immediately over the incubator cellar, is the brooder house proper, for in it are the hovers where the newly-hatched chicks from the incubators are placed. The remaining 68 feet is, strictly speaking, a nursery into which the chicks are removed as soon as they are old enough not to require any artificial heat beyond the ordinary temperature of the brooder house.

The floor joists of the brooder house are laid on concrete sills, filled in between with concrete on the plan known as "beam filling." Most brooder houses have the floor joists laid on wooden sills, which are always an invitation to the industrious rat. This little detail makes the brooder house at Sunny Slope Farm absolutely rat proof. The north and side walls of this house are built in the same way as those of the laying houses—that is, 2 x 4 studs are used, on both sides of which are nailed tongue-and-groove boards and heavy roofing paper, the joints of the latter being cemented. The roof is also made in the same way. This leaves an open air space of 4 inches in the walls and 10 inches in the roof, and makes the houses absolutely weatherproof as far as the north, east and west sides are concerned, except for the ventilating windows, which are described below and which are continually under control of the attendants.

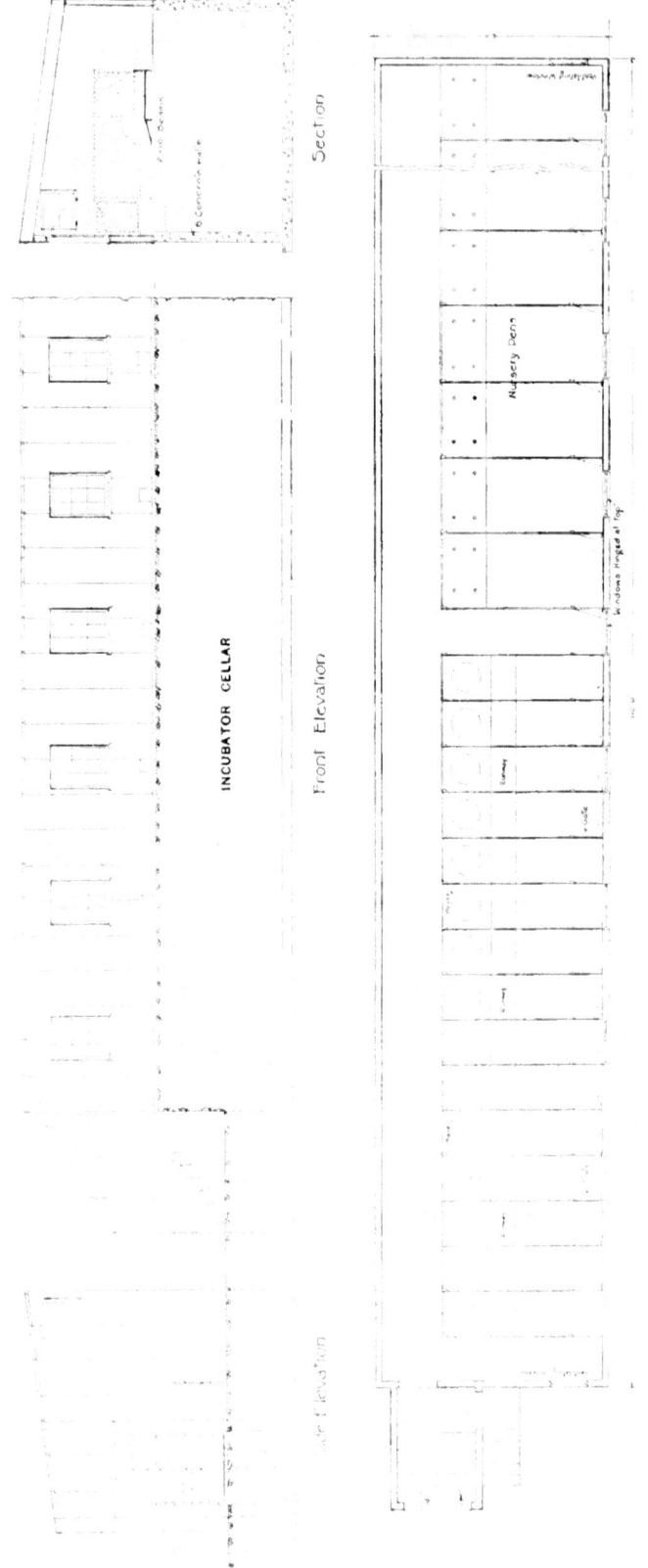

SHOWING ELEVATION, CROSS SECTIONS AND FLOOR PLAN OF THE INCUBATOR AND BROODER HOUSE

The front or south wall is built in the same manner, except that openings are made for the windows, which are of glass, and the little doors opening into the outside yards. Great care was taken in the construction of this building to have the windows and doors absolutely

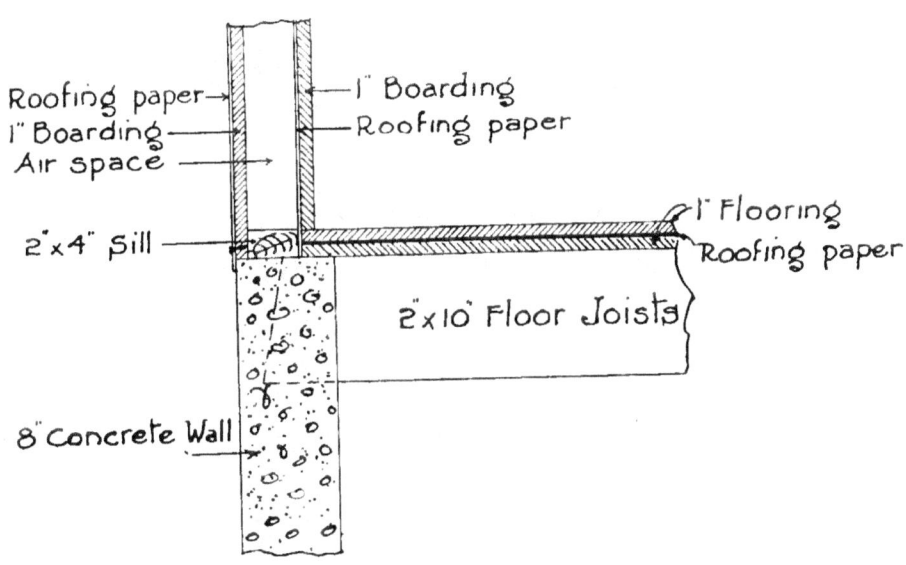

Detail of floor and wall construction of the Brooder House

draught proof, particularly the latter, and this has been accomplished. The little doors are built on the same principle as the door of a refrigerator or safe, so that when closed the outside air cannot find an entrance through them.

Along the back of the interior of this building runs a 4-foot passageway throughout its entire length. The brooder section proper is divided into fifteen runs of about 3 feet in width and 12 feet in length, which are separated from each other by a board 8 inches wide and an inch-mesh wire 4 feet high. In the back of these little runs is the hover, set on a false floor, 6 inches above the regular floor of the building. The lamp stands on the main floor, the chimney coming up through the false floor and going into the drum connection of the hover. This false floor is the width of the run and about three and one-third feet the other way. Hinged to it is a runway the entire width of the run, which gives an easy incline to the main floor of the run. This runway may be pulled up by a cord and pulley from the alleyway and used as a gate to hold the chicks right up to the hover when desired.

In each run there is a board next the alleyway, which when lifted up permits all the droppings from the hover to be swept into a box, which is also specially made the exact width of the run. The hover itself may be readily lifted up, as it is not in any way attached to the building or floor. This allows all work around the hover to be easily done from the alleyway.

To reach the front of the runs, and the windows in the front of the building, small gates have been put in the run fences near the front wall, which allows the attendant to pass down the entire south front from yard to yard.

The remaining 60 feet of the brooder house is divided into twelve pens, 5 feet in width, with the same sliding door adjustments separating them from the alley to make cleaning readily possible. These pens are the nursery where the chicks are introduced at three weeks of age, and where they learn to live without the brooder in the warm natural temperature of the room.

Much thought was given to the matter of ventilation in this house. If large, healthy, well-developed chicks are to be had, it is very essential

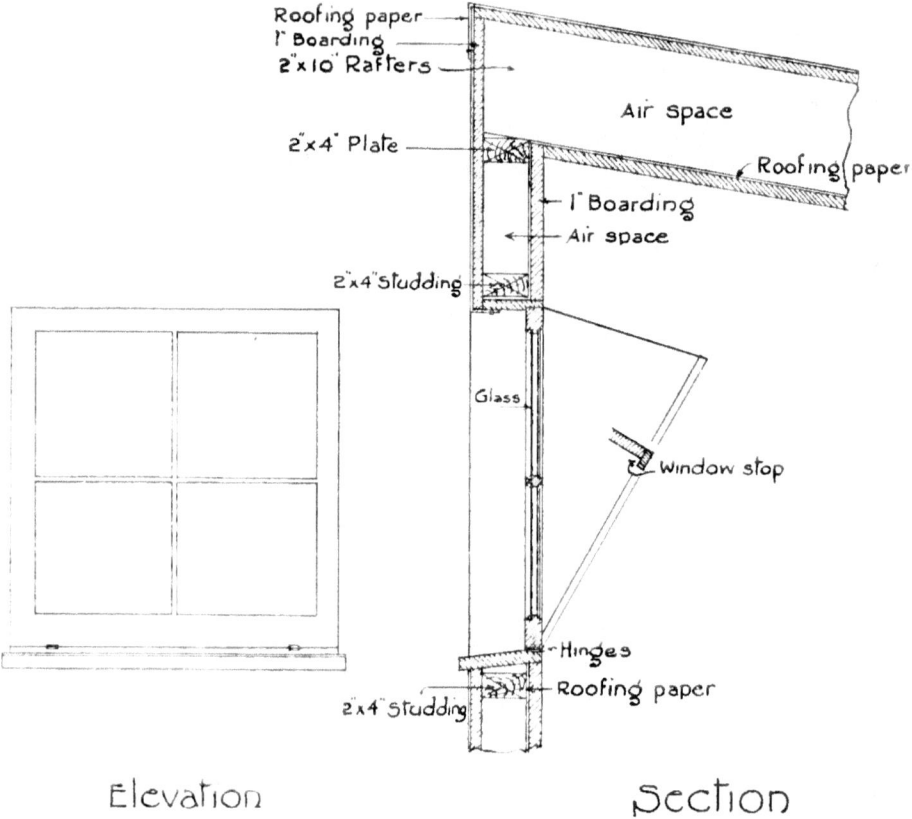

Detail of Ventilating Window of Brooder House

that their little lungs should be constantly fed from the moment they leave the incubator with an abundance of pure, fresh air. At the same time no draughts must be allowed to touch them.

The main ventilators in this house are in the highest point of the east and west corners of the building respectively. These ventilators are two windows about two and one-half feet square, hinged at the bottom and opening from the top. They are fitted with a set of cords and pulleys so arranged that an attendant can open them or close them at will from the alleyway. They have "V"-shaped solid sides, built out so as to enclose the windows when open in such a way that all air must be taken in or out from over the top, thus preventing any cold air striking down directly on the chicks. By keeping the windows on the non-weather side, or the side opposite to that from which the wind or storm is coming, open, the foul air is sucked over the top and out without creating a draught. The south windows through which the light is obtained are hinged at the top and open at the bottom in the same way as cathedral windows. They can be hooked out more or less as required.

Along the entire length of the house are run five 2-inch hot water pipes, placed alongside of the north wall about two feet from the floor. The hot water system is regulated so as to never allow the temperature in the brooder house to drop below 70° F. When the water indicator shows between 180 to 200° F. in ordinary April weather, the temperature in the house can be readily kept above that point.

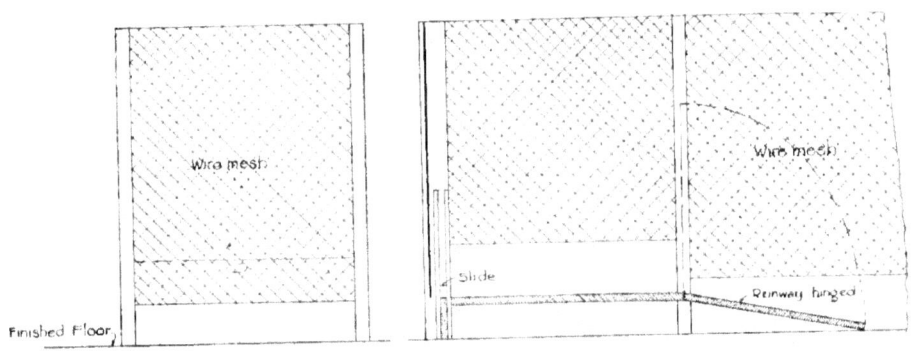

Details of wire divisions for Brooder Pens

In case the house gets too warm, the fire in the boiler is not dropped but the amount of fresh air is increased. It is realized that if healthy chicks are to be raised it must be done with pure air. In the Sunny Slope Farm brooder house no consideration is overlooked to reproduce summer conditions as nearly as possible.

Colony Houses

The pullets are brought to maturity in the colony houses, twenty of which were used on this farm last season. These houses have a floor space of 6 x 10 feet. They are 6 feet high in front and 4 feet at the back, with an ordinary shed roof, except that it does not project either in front or in the rear.

The framework is built on three skids. The outer ones of these are made of 3 x 4 studding, rounded at the ends in order that they may slide readily, and are 12 feet in length, projecting a foot at either end beyond the sides of the house. The centre skid is made of 2 x 4 studding.

These three skids are securely fastened together by four pieces of 2 x 4 studding. Across this is nailed the floor, which is made with inch stuff, tongue and grooved. The upright studs are made of 2 x 4 stuff. This framework is covered by matched boarding, and over the roof is placed heavy asphalt roofing, the joints of which are cemented.

In the front of the house is a door through which the attendant may enter, 2 feet wide and 5 feet high. On either side of this, and well up from the floor, are windows 45 x 27 inches. These windows are covered with medium-weight cotton duck and open outward. The advantage in having them open thus is that when open they act as an awning to exclude the sun from the coop and keep it much cooler than it other-

---

The colony system for growing stock is practiced quite extensively throughout New England on all the large poultry farms, and where such range can be had over rich grass land, it is a wonderful help in growing the youngsters. The objection, however, to most of these ranges is that there is too little natural shade for hot weather, and artificial arrangements must be provided. Professor Gowell was a great believer in this colony plan for developing pullets. The cockerels should be separated from the pullets as soon as sex can be distinguished.—EDITOR.

THE CORNING COLONY HOUSE

Where the young stock is kept on range to give it a sturdy constitution

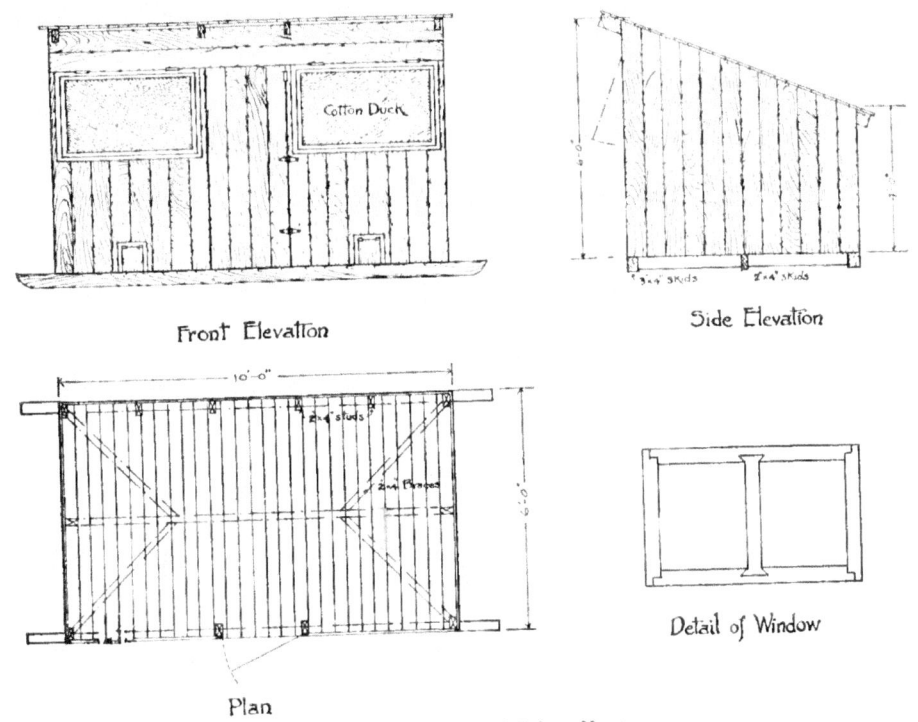

Plan and Elevations of Colony House

wise would be. Wire netting is also placed over these windows on the inside.

The pullets come in and out of this house through two small doors on either side of the main door.

With each four of these colony houses is provided a shelter 12 feet long and 9 feet wide, made of a framework covered with asphalt roofing. The front of this shelter, which faces north, is 3 feet high, and 2 feet at the rear.

Main Laying Houses

The main laying houses are 160 feet long and 16 feet wide, facing due south, without outdoor runs.

Theoretically, there would be no economy in having more than 1,500 birds in one flock, and this size house has been found abundantly large for this number of layers. True, this breaks all previous theories as to the floor space required for each bird, but in practice it has been found ample.

It is also better to have the building only 16 feet wide instead of 20, because the sun cannot be made to readily reach all parts unless the front elevation is made unnecessarily high. If the ceiling is high enough to allow the attendants to go through the houses without stooping, and it certainly should be, the sun has no difficulty in reaching to every foot of a house 16 feet wide. It is, therefore, economical to make it that width instead of narrower.

The foundation is placed on 8-foot cedar posts, placed 3 feet in the ground on a large rock or cement bottom. These posts are 8 feet apart, and are braced at the ends and cross tied at the corners. They are also

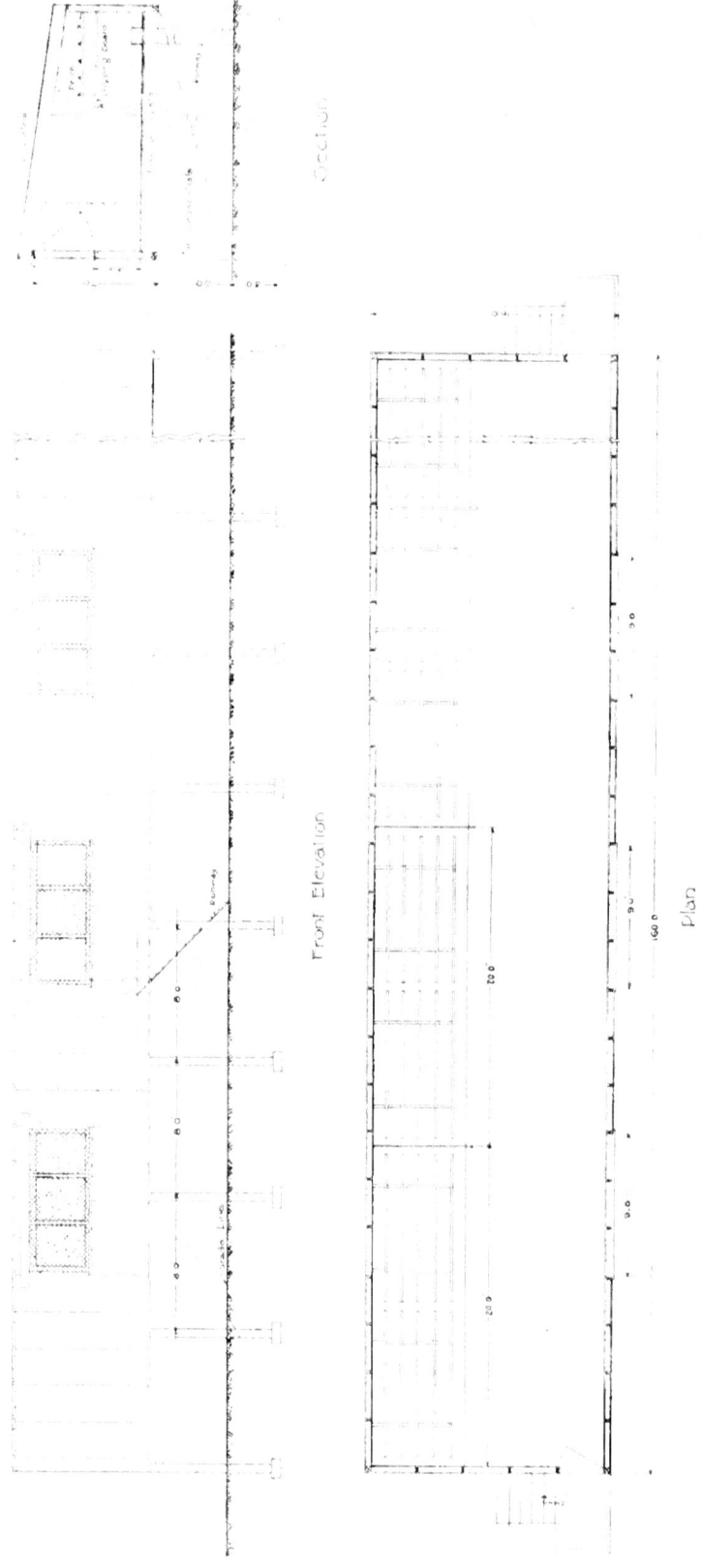

FRONT ELEVATION AND SECTION OF LAYING HOUSE. ALSO PLAN SHOWING THE ROOSTING PERCHES AND PARTITIONS FOR PREVENTING DRAFTS

The arrangement of this house is one of the secrets of the Corning's success. The interior is shown in the frontispiece and the exterior on the opposite page

Since this drawing was made the doors at the ends have been made six feet wide, consisting of two doors each three feet wide

braced both ways every 50 feet. This prevents the building from rocking. On top of these posts are placed 4 x 4 sills, and on these the framework rests.

The floor joists are 16 feet long, of 2 x 10 timbers, and are placed 3 feet apart. The elevation of the building is 7 feet in front and 5 feet behind. The uprights are 2 x 4 studs, placed 3 feet apart. At the corners the studs are doubled and spiked. The plates are also made of

The three Laying Houses. The interior of No. 2 is shown in the frontispiece

2 x 4 studs. The rafters are made of 2 x 10 studs and have no projection beyond the plates. This saves something in lumber, and makes it easier to make the walls at the back air tight.

The floor is doubled. The under floor is made of rough inch-boarding, running lengthwise of the house. On top of them is placed a covering of heavy asphalt roofing paper, every joint of which is carefully cemented. This prevents any draughts coming up through the floor. On top of this, laid crosswise of the under boards, is another floor of No. 4 matched tongue-and-grooved flooring.

This floor is not only absolutely air tight, but on account of its construction and the fact that it is 5 feet from the ground, it is proof against rats, skunks, weasels or vermin of any kind that prey on poultry or are a nuisance around a poultry house.

On the outside of the uprights, on the back and sides, inch-boards of good quality lumber are nailed. These are planed on the outside and covered with heavy two-ply asphalt roofing paper, well nailed down and carefully cemented at the joints. The nails have large galvanized heads, and are used so generously as to prevent bulging.

On the inner sides of the uprights another covering of heavy roofing paper is used, the joints carefully cemented, and over this is placed matched lumber. This gives an air-tight 4-inch vacuum, which is at

---

Professor Gowell, on the Gowell Farm, built his houses 20 feet wide, believing it was better economy than having them narrow, and, in Bulletin No. 144, says: "Nearly two years' use of this wide house shows its advantage over the narrower ones to be greater than was anticipated when it was planned. Its great width and the low-down door in the back wall make it much cooler in hot weather."

Sunny Slope Farm does not have outdoor runs, but the Maine Station has, and, especially during July and August, Professor Gowell said the fowls delight to go out into the yards early in the morning. On Sunny Slope Farm, however, open sheds are constructed under the house (something on the order of the plan sold by the Philo System), where the fowls can have a change of diversion—scratching in the loose soil.

No glass is used in the laying houses on Sunny Slope Farm, but at the Maine Station the front side of each section has two windows of twelve lights of 10 x 12 glass, screwed on, upright, 2 feet 8 inches from each end of the room. They are 3 feet above the floor. The space between the windows is 8 feet 10 inches long, and the top part of it down from the plate, 3½ feet, is not boarded, but left open to be covered by the cloth curtain when necessary.

Another change that Sunny Slope Farm has made in constructing the laying houses after the plan of the Maine Station, is that they have the house all in one room, placing 1,500 birds in one flock, while the Maine plan has the building divided by tight board partitions in twenty sections, each section being 20 feet long.—EDITOR.

the same time the warmest and coolest wall known, on the back and sides.

Under the floor, the back and sides are all boarded in and covered with paper, as in the upper part of the house, except that only a single covering of boards and paper is used.

Except for the window openings, the front wall of the house is constructed in the same way as the other walls.

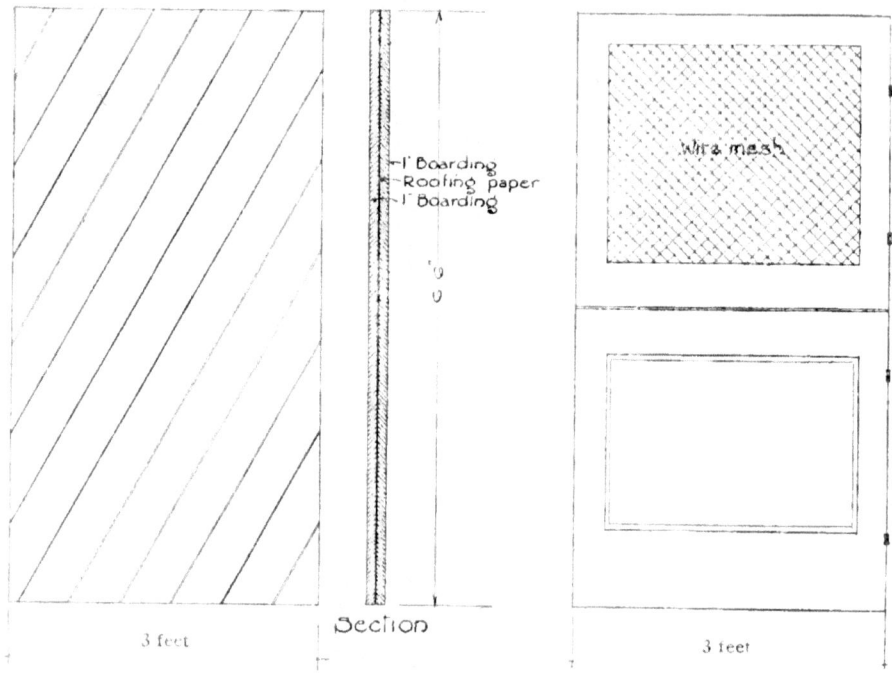

Details of inner and outer doors of Laying House. Two doors are provided, as the openings at each end of the house are six feet wide

The roof is made of seven-eighths sheathing, on top of which is a layer of two-ply asphalt roofing paper with all joints carefully cemented and well nailed down. The under side of the rafters is ceiled with a cheap grade of matched flooring.

Care is taken to have the paper lap at all the corners and joints and around the ridges, so as to prevent any draughts from getting through at these points.

There are eight windows on the south side of the house, and each of these is 3½ feet by 9 feet. They are 3 feet from the floor and run up to the plate. The object of this is to keep the draughts from blowing in on the floor, striking the fowls and scattering the litter.

The window frame is made of 4-inch stuff, seven-eighths of an inch in thickness, with two 4-inch supports placed so as to divide the window into three 3-foot sections. These come against the upright studs of the building, the latter not being cut out or weakened in any way to reduce the strength of the building.

This frame is covered with medium-weight cotton duck. Cheese cloth is too flimsy, and the duck lets in abundant air if care is taken to brush the accumulated dust off at regular intervals. Not a square inch of glass is used on the plant, outside of the brooder house and office and feed buildings.

A water shed 8 inches wide is built over each window, extending

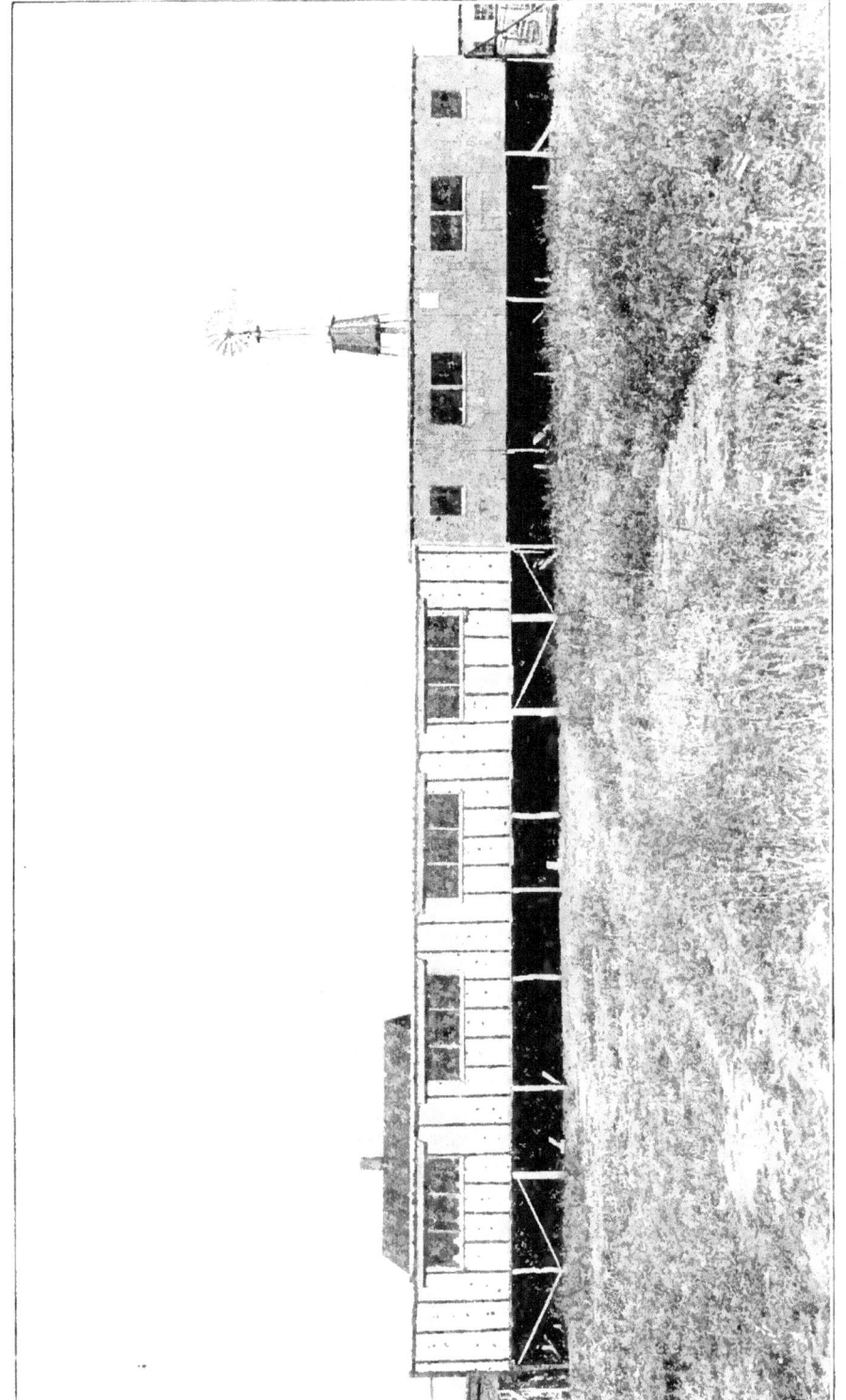

BREEDING HOUSES AT SUNNY SLOPE FARM

2 feet beyond the opening to prevent driving rains or storms from the south from beating into the house. The tar paper from the roof extends down over this storm shed.

The windows may be opened or closed as desired by being hinged at plate. They are kept open throughout the year, except on cold, blustery days and on winter nights.

Across the front of the window openings is placed 1-inch mesh wire netting to prevent the fowls flying out when the cotton-duck window is swung up.

The entrances to these houses are at each end. Seven steps lead to a platform 5 feet square, large enough to set feed pails on, which is surrounded on two sides with a hand rail. The doors are made of heavy matched lumber, of two thicknesses, with roofing paper between, the inner boards being nailed diagonally with those on the outside. The door is put together with one and one-half inch No. 12 screws. This makes the door very substantial. It is 6½ feet high and 3 feet wide.*

A second door is also provided, which opens outward, the main door opening in, the upper half of which has panel removed and is covered with wire mesh. At the bottom it is tight boarded. In milder weather the double door is opened, and the latter is used as a ventilator.

The roosting closets are built differently to those in any other house known to the writer. They are built in 20-foot sections, with close board partitions extending 12 inches beyond the dropping boards, which are 6 feet wide.

This extra width in the partitions is very important, as it obviates entirely the dangerous draughts which so baffled the late Professor Gowell in his large flock houses. When cross winds are blowing the fowls will go back into these closets to work and sun themselves.

The dropping boards are placed 3 feet above the floor, leaving plenty of room for the hens to work in the litter thereunder, and sufficient for the attendant to easily get under to gather the eggs that may be laid in the litter. A hen dearly loves to round out a nest in the straw litter and deposit her egg therein. It also gives the sun a chance to reach every corner of the floor from the front to the back at some period of the day.

There are two sets of roosts in each closet. These are made of five perches of 2 x 2 stuff, rounded at the top, nailed to crosspieces.

The first perch is placed 9 inches from the back wall, and the succeeding ones are 13 inches apart, measuring from centre to centre. This distance has to be regulated by the variety kept.

The crosspieces to which the roosts are nailed are securely hinged at the back, a foot up from the dropping board, and are hooked up to the ceiling when the dropping boards are being cleaned. They are supported by a leg a foot high in front, which keeps the perches up the required distance at roosting time.

There are openings to the yards under the house, which are placed under the roosting boards at the rear. There are five of these openings in each house. Boxes are built up around them a foot high to keep the litter in. A runway therefrom leads to the ground 5 feet below.

From the time the pullets are put into the laying house until warm weather in the spring comes they never leave the laying house. But they are as happy and contented as can be, and sing and lay eggs each livelong day.

In the late spring they are allowed the use of the space under the houses, but never beyond this.

The floor of the laying house is covered with a litter of at least

---

*Since the above was written, it has been found expedient, in practice, to increase the openings at each end of the laying and breeding houses, by substituting double doors for single, making the openings 6 feet wide instead of 3 feet.

This is a great improvement for ventilation in warm weather, and facilitates the handling of litter, in and out of the houses.

eight inches of wheat straw. When this litter should be renewed is a matter of individual judgment. At Sunny Slope Farm new litter is added from time to time as the straw becomes broken up, but the old is not removed, being left till spring. It has been found that the dust which accumulates from the breaking up of the litter supplies all required by the hens for a dust bath.

Nests are made of soap boxes and other boxes bought of nearby stores at a much less expense than it would be to have a carpenter make them. There are about three hundred boxes in each house. They are put up in four tiers and fill in the spaces between the windows, as shown in the frontispiece. Excelsior is placed in the nests.

The Breeding House

The breeding house is built exactly the same way as the laying house. The original house was 50 feet long, but in the spring of 1900 an addition 16 x 72 feet was added. Both buildings show in the illustration on page 32. The new building at the left of the picture. The combined houses give Sunny Slope Farm a capacity of over nine hundred breeders.

The small doors through which the fowls reach the ground are kept open every day, in order to allow the fowls free access to the open yards except on very wet or stormy days.

The Cockerel House

The house in which the young males are placed for fattening or bringing to maturity for breeders has a length of 30 feet and a width of 12 feet. It is built on posts, with the floor 5 feet from the ground and the superstructure put together in the same way as in the brooder and laying houses.

The dropping platform is also arranged in the same way as in the other houses, but the roosts are made 6 inches in width instead of only 2 inches. This is done to prevent the young birds from pressing on their breastbones, which are very tender in growing stock. In front of this house is a wire pen, 15 x 30 feet, covering over the top to keep out the sparrows.

The cockerels get into this yard through two small doors opening to a runway down underneath the house. Including the space under the house, the birds have a yard 25 x 30 feet. In the warm weather they spend a great deal of time in the shade and away from all draughts under the house.

The fences of these runs are made by putting in posts at suitable distances apart and over this is placed the wire netting. The boards upon the yards are used in order to keep the young males as ignorant as possible of what is going on in the world beyond the runs.

It is the rule on Sunny Slope Farm that the fowls must not be exposed to wet and stormy weather. Poultrymen are apt to be too careless in this matter. A fowl is just as miserable when exposed to drenching rains, or snowstorms, or heavy windstorms, as would be a human being, and an uncomfortable hen will receive a severe check in her laying. It is important that the hens be kept comfortable.—EDITOR.

Professor Gowell used vacated brooder houses for his cockerels. When the chickens reached the age of nine or ten weeks, and the cockerels weighed a pound and a quarter to a pound and a half, they were placed by themselves into vacated brooder houses, one hundred to a house. Each house has a yard in front, about twelve feet square.—EDITOR.

THE COCKEREL HOUSE

The Feedhouse and Workshop

This is a building two stories in height, with an area of 20 x 30 feet. At the present time the upper story is used as a dwelling by one of the attendants, the stairway leading thereto being from the outside. On the ground floor are the food bins for each variety of grain and meal used in the fowls' rations.

The machinery used in preparing the feed is all of the approved type. The gasoline engine is a 10 H. P. and furnishes the necessary energy for the bone cutter and the mash-mixing machine. The bone cutter was specially built for this plant, the ordinary sizes being too small to grind expeditiously the 175 pounds of bone used daily.

The mixing machine stands in close proximity to the bone cutter. This machine was also specially built for this plant, and is constructed on something the same lines as a cement mixer. The various ingredients of the mash are placed therein in the required proportion, and are mixed much more thoroughly than they could possibly be by hand, except in small quantities. The thorough mixing of the mash is regarded as one of the secrets of its effectiveness in producing large quantities of eggs.

A small shed, 12 x 16 feet, built of rough lumber, is conveniently placed. It is used for the storing of the droppings, and the side boards are placed on the inside of the upright studs in order that they may not be pushed off by the weight of the manure. If these droppings are sprinkled occasionally with sand or ashes there will be no odor from them.

The ice house is also one of the buildings on the plant, and is used for the preservation of the green bone until such times as it is required for food.

Operating the Incubators

The arrangement of the incubators in the cellar is worth some consideration. In purchasing the machines care must be taken to buy them built right and left. This will permit the lamps of the two machines being placed side by side and means a considerable economy in time in attending the lamps as well as floor space. They are placed side by side along the north and south sides and close to the wall. This arrangement leaves a good aisle through the centre for the operators. When installed they are carefully leveled with a spirit level.

Two tables are provided for convenience in turning the eggs. These are made three feet wide and a little longer than necessary to take two of the incubator trays side by side. They are built on revolving castors and their height is gauged so as to make them a trifle lower than the tray slides in the machine.

A great point made by the Sunny Slope Farm is the thorough mixing of the mash food. This is highly important, so that the different ingredients will be available to all the fowls. The workshop arrangement on this farm is very complete.—EDITOR.

About incubators, Professor Gowell says, in Farmers' Bulletin No. 357:
"There are many makes of incubators on the market, most of which will give fairly satisfactory results. The Maine Station has not tested many makes of incubators, and very likely some of the makes not tested would prove as satisfactory as the make used. Where many machines are used the hand turning of the eggs absorbs considerable time. Several turning devices are in vogue and equally good hatches have been obtained with them as when the eggs have been turned by hand. Machines that have artificial turning shelves will not hold quite so many eggs as when flat shelves are used, but the saving of time compensates for this.

Whatever make of incubator is used, pains should be taken to become thoroughly acquainted with the machine before the eggs are put into it. It is advisable for a person not familiar with the use of an incubator to run the machine empty for several days before filling it. After the eggs are put in, changes and adjustments should be made with the greatest care for fear of extreme results. By the use of an incubator it is possible to determine exactly the time when the chickens shall be hatched."—EDITOR.

THE FEEDHOUSE AND WORKSHOP

Showing bins containing feed in the rear and power meat cutter and mixer. Since the above photograph was taken a new 10 h. p. engine has been installed

By having two of these tables much time can be saved. While the eggs on one are being turned those on the other can be left to cool. By taking trays out of alternate machines the time for cooling can be increased as desired. Last season 5,000 eggs were placed in the incubators in this cellar, and from these 3,313 fluffy, bright, lively chicks were placed under the hovers in the brooding house.

On the leg of each machine is placed a tag, on which is provided blank spaces in which the following information is filled in:

1. The day and hour that the incubator is set.
2. The number of eggs placed in the machine.
3. The day, which will be the third day, when the eggs are turned for the first time.
4. The day, which will be the fourteenth day, when the eggs are tested out. Many incubator men test their eggs for the first time all the way from the third to the seventh day, but no test whatever is made on this farm until the fourteenth day. The principal reason given for testing on the seventh day, or earlier, is that the eggs taken out may be saved to boil for the newly-hatched chicks when they come. There is not enough in this, however, to take chances of losing valuable chicks thereby. In addition it saves time and labor.
5. The number of infertile, doubtful and fertile eggs found in the machine as a result of the test on the fourteenth day. The infertile eggs, of course, are removed from the machine and the doubtful ones are marked with a cross, so that when the hatch is off the operator can go over the remaining eggs. This enables him to perfect himself in the testing art.
6. The date, which will be the eighteenth day, when the incubator is closed and not interfered with again, until the doors are opened to remove the newly-hatched chicks.

Most people believe that an incubator does not begin to hatch until the twenty-first day. This is not always correct. The chicks begin to leave the shells on the twentieth day if the germs are strong and the temperature has been kept at proper height and the eggs were fresh when placed in the machine. If the temperature has been allowed to run low, the hatch is retarded and the chicks are apt to be not so strong as are those that come out on the twentieth day.

7. A final space is left in which is indicated the number of strong, healthy chicks hatched. Any weaklings are quickly disposed of.

This enables the operator to keep an accurate record of the work of his machine from season to season, and to correct any defects that may be observed.

Ventilation and Moisture of Incubator Cellar

It is impossible to properly ventilate many of the incubator cellars for the reason that the ceilings are too low. When a large number of incubators are being run there must be sufficient height to readily get the impure air from them out of the room. The windows in the hatching cellar at Sunny Slope Farm, which act as ventilators, are hinged at the bottom and drop into a "V"-shaped box with solid sides, forcing the air in or out, as desired, over the top, and not letting in cold gusts of air to strike directly on the machines.

Night and day these windows are kept at least slightly open, so that there is constantly a considerable change of air, insuring freshness. Fresh

---

Sunny Slope Farm has a greater belief in moisture than is general among poultrymen. Notwithstanding that the cellar on this farm is built upon a damp location, additional moisture is supplied by the pans mentioned above. This, too, despite the fact that the incubators used are of the "no-moisture" type, which goes to prove a theory the writer has held for years, that the "no moisture" claim is not founded on fact—that all depends upon surrounding conditions as to the amount of moisture to be used.—EDITOR.

INCUBATOR CELLAR

air in the machine is very helpful to the growing germ. The poisonous gases from the lamps must be driven out of the cellar or anæmic chicks will result every time.

The temperature is kept as nearly even as possible. It should never be allowed to go above 70° or below 40° F.

Moisture, and a large supply of it, is very important. In this cellar earthen pans are filled with water and placed on the floor almost directly below each pair of lamps, pushing them sufficiently far back against the wall as to be out of the way. If the atmosphere is particularly dry, after turning the eggs at night the concrete floor is thoroughly wet by sprinkling it with a watering can. This gives a relative humidity identical with that which occurs in natural incubation, being about 60°, and also provides the same amount of evaporation as in an egg under a hen.

This method of supplying moisture has proved most successful, for when the chicks are hatched the incubators will fairly run water, and it is no uncommon thing for the attendant to be obliged to take the hinges off the doors leading into the egg chamber as well as those through which the chick drawers are taken out.

When to Hatch

April and May are considered the best hatching months on Sunny Slope Farm, and the nearer the middle of April that the first hatches come off the better. Professor Gowell recommended April and May as the right time for hatching Plymouth Rocks. He knew a thing or two about bringing on young fowls, but in his writings he never gave the information the emphasis it deserved.

To be able to delay hatching until April and May has several advantages, among which are the greatly increased percentage of fertility in the eggs and the fact that when the chick is old enough to be moved into the colony house settled weather conditions are obtained.

The Corning Method of feeding is described in another chapter, which brings the pullet to maturity in about five months. Chicks hatched on this farm in the middle of April are laying by the middle of September and continue to do so throughout the ensuing ten months. If they are hatched earlier and forced along under this Method, they will commence laying earlier and are very liable to molt, which, with the Leghorns, means about two months' absence from the nests. It also necessitates the feeding of the pullets for two months, during which they are producing nothing.

With the heavier American breeds, such as the Plymouth Rocks and Wyandottes, maturity can be reached almost as quickly. We believe the reason Professor Gowell did not emphasize more strongly the hatching of chicks in April and May was because he knew the average poultryman would not give the care to his young flock nor force them along as he did. Under ordinary feeding methods it takes from seven to eight months to mature a bird of either the American or Mediterranean breeds and get them started laying.

A pullet rarely begins to lay in very cold weather, and if a supply of winter eggs are to be had she must be started to work on the nests before the thermometer goes regularly below freezing at night. If she

---

Professor Gowell's main idea in having May hatches is that they yielded more chicks, there being better fertility, and natural brooding conditions are better. In Bulletin No. 130, referring to tests made at the Maine Station, in order to study the hatchability of the eggs from the same lot of hens through their first laying year, a pen of fifty pullets was set apart for the purpose. They were hatched late in May and commenced laying in October, continuing laying moderately through November and December. The fifty birds were mated in November with two cockerels that did not quarrel, and these matings continued through the ten months' test.—EDITOR.

is hatched late she must be forced along quickly to the laying point to have any returns from her in the winter months.

On Sunny Slope Farm one of the problems to be solved has been to so hatch and rear the pullets as to bring them to the laying point, in flocks of about 1,500—before the yearling hens have gotten well into the molt and stopped laying. To accomplish this it has been found advisable to have the brooder house well filled with chicks by March 20th. This produces a large flock of pullets ready for the laying house by the last of August, and they should be brought to this stage without undue forcing, which will minimize the danger of their molting.

It is true that often a small percentage will molt—but it has been found that by mid-September such a flock will be laying strongly, and the few which do molt will produce eggs until well along in December. During this period these pullets have given a large number of eggs—through October, thus bridging the time between the coming in of April and May hatches and the going out of the yearling hens.

Taking Care of the Chicks

The hovers are operated with the same care and in the same methodical way as the incubators. A tag is placed on the wall of the alleyway back of each brooder yard, and on this is indicated the number of chicks, the day and the hour they were placed in the brooder and the losses from any cause as they occur. These tags are also filed away from season to season for reference, and they form a valuable guide as to what can be banked upon. Less than ten per cent. of all chicks placed in the brooder house were lost from all causes last season.

On account of lack of room in the brooder house a number of chicks had to be removed when they were between three and four weeks old to the colony houses to make room for new hatches from the incubators. Among these there was a considerable mortality, as they were not sufficiently feathered to provide the necessary warmth in the cold nights. The total loss of birds in the colony house from hawks and other causes was about 400. There were 899 cockerels raised to sufficient size for broilers or were matured for breeders. The pullets placed in the laying houses numbered 1,953. Of these 453 were placed in what is known as No. 1 house and 1,500 in No. 2 house.

It is a fixed belief at Sunny Slope Farm that the chicks should be handled as little as possible; so when a change is being made from one place to another it is accomplished by removing the sliding board, opening into the alleyway, which is then blocked so that the chicks can only go in the one desired direction. The board opening into the nursery pen to be occupied is raised, then the attendant walks behind these chicks and quietly and easily moves them along until they are in their new quarters. When removing them to the colony house a box is placed at the little door leading from the house to the yard. In this box is another door corresponding to that in the brooder house, and the front of the box is made of wire mesh. The chicks are quietly driven into the box and when it is comfortably filled the door is closed and they are carried down to the colony house, which is to be their home until they are sufficiently matured to take their places in the laying house. To facilitate the emptying of the boxes just described, doors are placed at either end.

Records shown the writer, upon his visit to Sunny Slope Farm, gave a very small loss in the brooders. In the spring of 1909 7,505 chicks were hatched at Sunny Slope Farm. There were 1,102 deaths in the brooder house, and up to August 15, 1909, 481 deaths on range from crows, rats, etc.—EDITOR.

Feeding Newly-Hatched Chicks

The chicks hatched at Sunny Slope Farm are not fed for forty-eight hours after they come out of the shell. The last act which the chick performs before breaking the shell is to absorb the yolk, which makes food unnecessary for at least two days.

The third day after hatching, the chicks are fed every two hours, of a good commercial chick food, not heavily, but just enough to enable them to readily fill up their crops. The feed is placed on the floor in close proximity to the hover. The fifth day after hatching they are allowed to run in the little hover yard and then their feed is thrown to them in litter. Small sized grit is mixed with the chick feed.

This litter is made of wheat chaff, or the screenings taken out of

---

Professor Gowell's method of feeding newly-hatched chicks is as follows (Bulletin No. 130):

"The best method of feeding young chicks is at present a matter of some uncertainty. Many different kinds of food and different ways of feeding give good results.

One condition appears to be imperative, and that is, that the young things, until they are at least three weeks old, be not allowed to overeat. We have guarded against this by watching them closely and examining their crops for emptiness just before feeding time. This enables them to eat four good meals a day and be hungry at feeding time. Where regular full meals are given they are allowed at the troughs only a short time. A long-drawn-out meal to enable them to clean up the dishes impairs their digestion, and ruin follows.

Where small broken grains and meals are kept constantly within reach of the young things, either in the litter or small troughs, the crops never appear to be empty, neither are they ever crammed full, as they are when fed at regular hours, and yet the birds live well and seem to thrive when they are within easy reach of food all of the time.

At the present time the Station is studying young chick feeding closely, for it is the most difficult feature of the whole poultry industry. We can now give no better method than that practiced in raising the chicks during this and the last season, because by it few birds have been lost and good thrift has been secured.

Infertile eggs are boiled for half an hour and then ground in an ordinary meat chopper, shells included, and mixed with about six times their bulk of rolled oats, by rubbing both together. This mixture is the feed for two or three days until the little things have learned how to eat. It is fed sparingly, in the litter and sand on the brooder floor.

About the third day they are fed a mixture of hard, fine broken grains, i. e., cracked corn, wheat, millet and pinhead oats, as soon as the birds can see to eat in the mornings. This is fed in the litter, care being taken to limit the quantity so they shall be hungry at ten o'clock. Several of the prepared dry chick foods have been tested. They are satisfactory when made of good, clean grains without grit. The grit and charcoal can be supplied at less cost and must be freely provided.

At ten o'clock the rolled oats and egg mixture is fed in tin plates with low rims. After they have had the food before them five minutes the dishes are removed and they have nothing to lunch on, except a little of the fine broken grain which they scratch for. At one o'clock the hard grains are again fed, as in the morning, and at four-thirty to five o'clock they are fed on the rolled oats and egg mixture, giving all they will eat until dark.

When they are about three weeks old the rolled oats and egg mixture is gradually displaced by a mixture made up of two parts, by weight, of good clean bran, four parts corn meal, two parts middlings or red dog flour, one part linseed meal and two parts screened beef scrap. This mixture is moistened just enough with water so that it is not sticky, but will crumble, when a handful is squeezed and then released. The birds are developed far enough by this time so that the tin plates are discarded for light flat troughs with low sides.

The hard broken grains may be safely used all the way along and the fine meals left out, but the chicks do not grow so fast as when the mash is fed. There seems to be least danger from bowel looseness when the dry grains only are fed, and it is very essential that the mash be dry enough to crumble, in order to avoid that difficulty. Young chicks like the moist mash better than though it was not moistened, and will eat more of it. There is no danger from the free use of the properly made mash, twice a day, and being already ground the young birds can eat and digest more of it than when the food is all coarse. This is a very important fact and should be taken advantage of at the time when the young things are most susceptible to rapid growth. But the development must be moderate during the first few weeks. The digestive organs must be kept in normal condition by the partial use of hard foods, and the gizzard must not be deprived of its legitimate work and allowed to become weak by disuse.

By the time the chicks are five or six weeks old the small broken grains are discontinued and the two litter feeds are wholly of screened cracked corn and whole wheat. Only good clean wheat, that is not sour or musty, should be used."—EDITOR.

INTERIOR OF THE CORNING BROODER HOUSE

the bottom of the hay mow. If neither of these is available, straw is cut fine with a clover cutter and is used as a substitute. The litter should be put in deep in these yards—at least a couple of inches of it.

While the bulk of the feed is fed in this litter, a small portion is scattered around the hover where no litter is kept, so that any weak chick may get it without too hard work.

From the start the chicks have water before them, placed in sanitary drinking fountains which are thoroughly washed and refilled three times a day—first thing in the morning, at eleven o'clock and again at three-thirty o'clock. These hours are not set arbitrarily, but were decided upon from the fact that it is at these hours approximately that the chicks drink the greatest quantity.

When the chicks begin to work in the litter, they are fed but three times a day instead of every two hours. While not overfeeding, enough grain is thrown into the litter so that the little fellows in scratching will always find something to reward them.

About this time beef scraps are added to the ration, and care is taken that these are ground fine. This is fed at noon. To each 100 chicks at the start a couple of little piles are thrown into the litter. Each pile contains about one handful. The amount of beef scrap is increased day by day until six handfuls are fed to each pen of 100 chicks.

The chicks are fed in this way until they are six weeks old. Then they are given wheat and cracked corn. This change is made gradually by mixing with the chick feed, each successive day feeding less of the latter until it is taken out altogether.

Feeding Pullets While on Range

At Sunny Slope Farm the pullets are moved from the brooder house to colony houses and are given free range when they are six weeks old. From that time until they are mature enough to be placed in the laying houses they are made to do a lot of hustling for their living. Some poultry breeders throw corn to the pullets in the vicinity of their quarters early in the morning. This is a serious error. At this stage in a chick's life it is greatly to its advantage in developing stamina and hardihood to make it hustle for its food. The pullets also enjoy ranging, as can be seen by watching them playing tag with each other and chasing the worms and insects.

In Bulletin No. 130 Professor Gowell says:
"When the cockerels are taken out for finishing, the pullets of the same age are moved to the grassy range, still occupying the same portable houses in which they were raised. At this time the method of feeding is changed, and dry food is kept by them constantly, in troughs with slatted sides and broad detachable roofs, so it may not be soiled or wasted. The troughs are from 6 to 10 feet long, with the sides 5 inches high. The lath slats are 2 inches apart and the troughs are 16 inches high from floor to roof. The roofs project about 2 inches at the sides and effectually keep out the rain, except when high winds prevail.

The roof is easily removed by lifting one end and sliding it endwise on the opposite gable end on which it rests. The trough can then be filled and the roof drawn back into place without lifting it. This arrangement is the best thus far found, for saving food from waste and keeping it in good condition. When dry mash is used in it there may be considerable waste by the finer parts being blown away. When used for that purpose it is necessary to put it in a sheltered place out of the high winds.

In separate compartments of the troughs, they are given cracked corn, wheat, oats, dry meal mixture, grit, dry cracked bone, oyster shell and charcoal. The dry meal mixture is of the same composition as that fed to the laying hens. The troughs are located about the field in sufficient numbers to fully accommodate all of the birds.

The results of this method of feeding are satisfactory. The labor of feeding is far less than that required by any other method followed. The birds do not hang around the troughs and overeat, but help themselves, a little at a time, and range off, hunting, or playing and coming back again, when so inclined, to the food supply at the troughs. There is no rushing, or crowding about the attendant, as is usual at feeding time where large numbers are kept together."—EDITOR.

In the mornings the grass is thick with insects and there are always plenty of pickings to be had if the pullets are forced to look therefor. To satisfy their appetites they are kept busy until about eleven o'clock, when they come into the shade for rest. It is late in the afternoon before they are given any grain, when it is scattered on the ground near the colony houses. This is a mixture of two-thirds whole wheat and one-third cracked corn. At this time they are given all of this that they will pick up clean.

Inside each colony house is a feeding trough for mash, and this is supplied in such quantities that it is never entirely consumed. Every afternoon at three o'clock fresh mash is placed in the trough, and whatever little may be left over from the day before, is mixed therewith.

To provide a sufficient amount of grit two basins are placed at each colony house and are always kept filled. Care is always taken to use a grit which carries a large percentage of lime. This helps to make bone, and the pullets come into laying without causing any trouble. The pullets also have a wonderful affinity for hard coal ashes, and large quantities of them are regularly provided. It is surprising the amount of ashes the pullets will consume.

This system of feeding sends pullets to their quarters with full crops, which is very essential if rapid growth is to be had.

There stands in each house an automatic drinking fountain, which holds five gallons of water, so that it has to be filled only every other day. This affords a considerable saving in labor.

Feeding Laying Pullets

When the pullets are put into the laying houses they still receive their main feed of grain at night. Six quarts of wheat and corn, varied in proportion according to the weather, for each hundred pullets are scattered in the litter an hour before sunset on clear days, and fifteen minutes earlier on cloudy days. The litter is at least eight inches deep and preferably of wheat straw. The grain is thrown on top of this, and being fed in this quantity the pullets are able to fill up quite easily at just the time when you want to get their crops chock-full. As she moves and scratches, the pullet buries the remaining grain in the litter. When she leaves the roost in the morning she has to work like a beaver to get out the remaining grain, which gives her the needed exercise and starts her blood well in motion for the day.

By thus feeding the extra quantity at night the attendants are saved the necessity of another trip with the morning ration of grain, and the burying of it in the litter. It is necessary that this feed more than any other should be so fed as to make the fowls work hard for it, and con-

---

Professor Gowell's method of feeding laying stock and that followed by Sunny Slope Farm are practically the same. There is a difference in the Gowell mash, as follows:

Wheat bran ....2 parts
Corn meal ....1 part
Wheat middlings ....1 part
Linseed meal ....1 part
Gluten meal ....1 part
Beef scraps ....1 part

The mash contained one-fourth of its bulk of clover leaves and heads obtained from the cattle barn. The clover was covered with hot water and allowed to stand for three or four hours. The mash was made quite dry, and rubbed down with the shovel in mixing, so that pieces of clover were separated and covered with the meal. This mash was fed to hens in artificially heated houses where the temperature was always above freezing. This method of feeding green food in mash was abandoned by Professor Gowell some years after, and clover cut in half-inch lengths was fed dry, separately.—EDITOR.

sequently it must be buried deep in the litter. At eleven o'clock a small quantity of oats is fed in the litter (two quarts for each hundred hens). In very cold weather a little buckwheat is mixed with the oats on alternate days. Buckwheat is too fattening to have a part in the daily ration.

The mash troughs which are placed under the dropping boards, two troughs being provided for each 20-foot section, are filled with mash, twenty-two pounds of mash being given to each 200 hens, at three o'clock in the afternoon. This mash is made by thoroughly mixing the following ingredients in the proportions named:

Wheat bran .... 8 parts
Ground oats (not fine) .... 4 parts
Wheat middlings .... 1 part
Old process oil meal .... 1 part
Gluten meal (highest quality) .... 1 part
Corn meal .... 1 part
Cut green bone .... 16 parts

This mash must be thoroughly mixed, so as to have the juices from the animal food taken up entirely by the ground grain. Absolutely no water, only the animal juices in the cut green bone is used in making this mash.

It is sweet-smelling and palatable enough looking to tempt any man. When ready for the hens there is not the slightest appearance of the green bone in it, all these particles having been thoroughly covered by the adherence of the meals.

Where it is not convenient to cut green bone, beef scraps may be substituted. Where small flocks are being handled, the mash can readily be mixed in a tub—or large pan—with a wooden paddle. This method was practiced at Sunny Slope Farm until the flock became so large that labor-saving machinery became an absolute necessity.

Feeding Cockerels for Broilers

At six weeks of age, as a rule, the cockerels are transferred from the brooding house to the fattening pen, or just as soon as their sex can be surely determined. The Leghorn cockerel is very precocious and develops the masculine traits at a much earlier age than those of American or heavier breeds.

The first thing in the morning the cockerels are fed a mixture of grain, composed of two parts of cracked corn and one of wheat. This is thrown on the floor, on which no litter is placed, as it is desirable that they should do as little moving about as possible. They are given all they will eat up clean.

Bulletin No. 90 of the United States Department of Agriculture states:

"A very large proportion of the cockerels raised in New England are sent to the market alive, without being fattened. Quite extended experiments at the Maine Station with many birds, in different years, indicate very clearly that keeping the cockerels for a few weeks with special feeding will add materially to the selling price. Not infrequently this will make the difference between loss from the low price obtained for slow-selling unfattened birds and the profit from comparatively quick-selling specially fed birds at a much higher price. The higher price is due partly to the increased weight and partly to the superior quality of the well-covered soft-fleshed chickens. As the bulletins containing the results of these feeding experiments with cockerels are out of print, the following brief summary of the results obtained is given:

The number of pounds of grain required to produce one pound of gain in fattening cockerels was ascertained in experiments comparing the effect of housing, the effect of age, and the effect of skim milk. The grain mixture used in these series of experiments was the same, consisting of 100 pounds of corn meal, 100 pounds of wheat middlings and 40 pounds of meat meal. This was fed as a porridge thick enough to drop but not to run from a spoon."—EDITOR.

FLOCK OF COCKERELS AT SUNNY SLOPE FARM

At ten-thirty o'clock they are given a bountiful supply of green food. At three-thirty o'clock the mash boxes are filled full and they are given all of this they will eat. This mash is made in the same way as that described for the laying pullets, except that the proportion of corn meal is very considerably increased.

This method of feeding has been found superior to any of those which use large quantities of milk in the mash, as it produces a broiler the meat of which is sweeter and more juicy. It has also been found more effective in pushing them to maturity.

Feeding the Breeding Stock

The birds in the breeding house are fed in exactly the same manner as those in the laying houses. That this method is correct is amply proved by the fact that in the past two seasons, since it has been adopted, the fertility of the eggs has averaged 90 per cent. or better, and the germs have been exceptionally strong. The chicks from these hens have great vigor and vitality, and grow rapidly. It keeps the males as well as the females in a strong, healthy condition, and does not make them too fat.

To secure a heavy supply of eggs the hens must be well fed, but to have a high percentage of fertility it is important that they should not be allowed to become overfat.

Feeding Hens Through Molt

At Sunny Slope Farm no hens are carried through the molt except those required for breeders. As soon as the pullets finish their first laying season, which lasts approximately ten months, they are sold at once for breeders. There is always a demand for these birds.

Care must be taken in feeding a molting hen not to let her take on fat. At this period in her life a hen is much less active than when she is laying, and is much given to "standin' 'round." It is therefore necessary to see that all the grain is buried deeply in the litter. The amount of cracked corn given is materially lessened and the quantity of mash feed is cut down by at least one-half.

If the hen is going to feather well and keep her strength some animal food is necessary, but she does not require it in so large quantities as when being fed for eggs. With the exceptions noted, molting hens are fed the same as layers.

When the hen begins to lay, the amount of mash is increased to the requirements of the hen, gradually, until it reaches the point of the laying ration.

Care is exercised on Sunny Slope Farm that the hens are kept in good condition without becoming overfat. Like the layers, they are kept active. The high percentage of fertility proves that their method in this particular is correct.—EDITOR.

Molting is a period that is a great strain upon the vitality of the fowls, and in order to carry them through safely it is necessary that their feed be of a more stimulating nature than when they are in laying condition. This fact is plainly exemplified at Sunny Slope Farm.—EDITOR.

Mash—Morning or Night?

It is a much disputed question whether mash should be fed in the morning or at night, but on Sunny Slope Farm it has been definitely answered to the satisfaction of its proprietors.

Professor Gowell used to feed his layers twice a day only, and kept the mash always before them. On Sunny Slope Farm better results have been obtained by feeding the mash in the late afternoon.

To keep her body in a perfectly healthy and natural condition, the hen must spend the greater part of the day in activity. If she is permitted to cram her crop with mash in the early morning she will lie around until the middle of the day or later, in a semi-dopy, sluggish condition.

The theory of the "morning mash" poultrymen is that by giving the hen warm mash in the morning it heats her blood and makes her more comfortable. This is a fallacy. It does not send the blood coursing through her veins nor make her nearly so comfortable and happy as to be forced to hustle, and hustle hard, for her morning meal.

If a hen is going to lay well she must be sent to roost with a full crop. For this reason, on Sunny Slope Farm she is given the mash late in the afternoon, and this is followed a short time before she goes to roost with an abundant feed of grain. Before she starts to pick the grain at all her crop is full of mash, but there are always a number of small cavities into which the hard grain can be pushed, which puts her to bed with a full crop of egg-making material—"It works while she sleeps."

Fresh Cut Bone

Fresh cut bone is a valuable egg producer, and an average of more than an ounce a day is fed to the layers and breeders at Sunny Slope Farm. It forms half of the mash mixture, and the hens are given all of this they will eat at one feeding.

Green bone is cheap in price and highly nutritious, easily digested and heartily relished by the fowls. It is stimulating to the egg-producing organs, but more in the way of strengthening than simply inciting to greater activity. The feeding of it is not followed by reactionary results, as is the case when condiments are used.

It not only imparts strength to the egg organs, but it contains in about equal proportions the same elements as the egg. Consequently, it is a valuable food. It has a noticeably favorable effect upon the fertility and hatchability of the eggs and upon the chicks after they are hatched.

Great care must be exercised in the selection of bones and meat put into the cutting machine. Unless this is done it is a very simple matter

---

On this point, in Bulletin No. 90, Professor Gowell says: "Years ago the 'morning mash,' which was regarded as necessary to 'warm up the cold hen' so she could lay that day, was given up, and the mash was fed at night. The birds for several years prior to 1903 were fed daily throughout the year as follows: Each pen of twenty-two received one pint of wheat in the deep litter early in the morning. At 9.30 A. M. one-half pint of oats was fed in the same way. At 1 P. M. one-half pint of cracked corn was given in the litter as before. At 3 P. M. in winter and 4 P. M. in summer they were given all the mash they would eat up clean in half an hour."—EDITOR.

In all the Maine Station feeding formulas Professor Gowell does not include green cut bone. While the writer has never read or heard of his opinion on this article of food, it is known that Professor Gowell personally preferred beef scraps. Green bone should be fed the day it is cut, and care should be taken that no tainted bone or meat is used. This is a hard matter to avoid where a large amount of bone daily is needed.—EDITOR.

to give the fowls an aggravated dose of diarrhea, and diarrhea caused in this way is practically incurable. From a financial standpoint, it is much better to kill the fowl than to try to doctor her back to health.

The bone and meat scrap as it is supplied by the butcher is gone over piece by piece, and all the salt or putrid pieces are thrown to one side. No chance is taken of allowing any meat that has been in brine to get into the hen's mash.

The attendant soon becomes quite expert in selecting the fresh from the salt bones. If the flavor of the eggs is to be maintained, care must be taken that no tainted bone or meat are used.

The bone is weighed out in exact proportion in preparing it for the mash, and as much bone is used as all other ingredients in the mash combined. This gives the laying stock something more than an ounce per head each day.

Green Food

All the birds on Sunny Slope Farm are given an abundance of green food, the supply starting when they are three days old and continuing until they are finally disposed of.

The little chicks are given grass and clover cut fresh every morning and reduced to lengths of about one-eighth of an inch. This is thrown into the litter in abundant quantities and they are allowed to eat it at will throughout the day. As soon as the chicks are removed from the brooder house they are placed on range, and of course get their own supply of green food.

When the pullets are placed in the laying house the green food has to be supplied to them. This is obtained by sowing a sufficiently large patch of ground with winter wheat in the late summer. This comes up very readily and is cut and fed in large quantities every morning, at least a bushel basket packed down being given to each 200 hens.

When this wheat gets covered by snow so that it is not cutable, green food is supplied by feeding short cut clover or alfalfa which has previously been gathered and cured. This is prepared for the fowls daily. It is placed in large tubs and over this is poured boiling water through a watering can. It has been found that the clover takes up water much more evenly when it is thus sprinkled on than when it is poured on. Sufficient water is used to thoroughly moisten it.

Then the tub is covered with burlap or old sacking and allowed to stand for thirty minutes. This limit of time is very important, for the reason that clover or alfalfa becomes brownish in color as well as soggy if allowed to steep longer, and it is not nearly so palatable to the hens.

The quantity of water used in making this green food mixture is not always the same, for sometimes the clover will suck up a good deal more moisture than it does at other times. The weight of the food is always more than doubled by the addition of the water.

The yards of the breeding pens are plowed as soon as they can be worked in the spring and are sown heavily with oats. The oats grow very quickly and after they have a fair start they will beat the hens, what is eaten down during the day being fully made up by the growth over night.

In the winter and early spring months the breeders are supplied with green food in the same way as the layers. A large amount of green food fed to the breeding stock adds greatly to the strength and fertility of

---

Professor Gowell was a firm advocate of clover. He said: "Poultry keepers do not begin to realize how valuable a food we have in clover." Clover supplies the much-needed mineral elements (ash) so necessary to the vigor of the new-hatched chick, and that much-desired mineral element must be in the egg from which the chick is hatched.—EDITOR.

FLOCK OF BREEDING COCKERELS

the eggs. To the stock in confinement in the summer and to all the birds in the winter and early spring months the green food is fed to them warm at nine o'clock each morning.

Much has been written about sprouted, or "processed" oats, but the whole matter is very simple. A frame made of flooring, or any boards 4 or 5 inches wide, and set upright, gives the growing bed.

Roofing paper, laid on level earth, makes a good bottom; concrete or board floors are equally good. In any event, the water should be allowed to drain away.

Frames 3 x 6 feet are used at Sunny Slope Farm, but any desired dimensions can be made.

Dry oats are spread in the frames, not over one inch in depth, and are *thoroughly* and evenly sprinkled with cold water every day, for ten to fourteen days, when the green sprouts are about six inches long.

A dark, cool cellar is best adapted for this operation, though sheds or other buildings can be utilized.

Oftentimes, as the oats swell and sprout, there will be upheaving spots, or islands, showing an uneven surface over the bed. These "islands" should be most thoroughly soaked with water, which will in a day or two bring the whole bed to a level growth. In bulk, this process produces a full four parts for one part planted, and makes a most excellent green and oat food. Better results are obtained by following this plan than other methods.

Drinking Water

Chemical analysis shows that more than three-fourths of an egg is composed of water. It is therefore essential that the fowls should be given an abundant supply of water, that it should be pure, and placed before them in such a way as to prevent their fouling it. At Sunny Slope Farm the water is given in automatic fountains which hold about five gallons apiece. A sufficient number of these are placed in each house to meet the requirements of the day. In the laying and breeding houses this is given in the morning, and in the cold months it is put in the fountains boiling hot.

This meets all the advantages claimed by the advocates of warm morning mash in heating up the fowl's system if it has been chilled through the night. It has the same effect upon the bird as a cup of hot tea or coffee has on a man whose system has become chilled from exposure or other causes. Together with the work the hen has to do to dig her breakfast out of the litter, it sends the blood circulating rapidly through her veins and makes her active and lively almost as soon as she is off the roost.

If cold water is given, the pullet will stand around dumpy, often for a couple of hours after leaving the roost in the morning, and much of the advantage that has been obtained by discarding the morning mash will be lost.

After a little experience the amount of water that the hens require can be readily gauged, so that the fountains are practically empty at night.

Young chicks are given water simultaneously with their first food, and plenty of it is always before them. Water is kept in each of the colony houses for young stock and replenished just as often as the fountains are empty—about every other day.

Professor Gowell considered water one of the greatest, if not the greatest, "egg foods" that could be given hens. In his class in the University he urged a constant examination of the drinking fountains, that they not only are filled, but that the water in them is fresh and clean.—EDITOR.

Charcoal, Grit and Oyster Shell

Charcoal is kept regularly before the fowls. It is fed in automatic hoppers, which are filled once a week. It does not affect all breeds alike, but it seems to make the Leghorns on this farm susceptible to colds, and for this reason it is not kept constantly before them, unless a hopperful lasts them a week. The hens are very fond of it, however, and this supply usually is consumed in a couple of days.

Charcoal has no equal as a bowel regulator, and it purifies the crop and keeps it sweet. Only coarse charcoal is used, as there is less waste in it and the fowls seem to prefer it.

The hens have free access to grit all the time, and care is taken to secure a grit that is really sharp, that does not crumble, and that carries a stiff percentage of lime. Do not make the mistake that a limestone is desirable.

Coarse oyster shell, perfectly free from dust and fine particles, is also kept at all times in front of the fowls. If there is a good percentage of lime in the grit used, less oyster shell will be required. Growing birds also need it for bone-making material. Good shells cannot be had on the eggs unless plenty of lime is supplied, and a good shell adds materially to the appearance of the eggs.

Hard Coal Ashes

The affinity existing between a chicken and hard coal ashes from the time the chick begins to eat until its final passing is remarkable. The experiments at Sunny Slope Farm, with hard coal ashes, have been on the increase from year to year.

Ashes were first placed in heaps on the colony range, and disappeared like frost before the sun.

Next they were placed at the end of the chick runs, and the "yellow babies" made way with them, greatly to their own benefit. Charcoal was then abandoned in the laying houses and sifted ashes, in feed hoppers, were substituted, resulting in a marked saving in grit and oyster shell. The necessity of keeping the brooder house runs in a clean, dry and sanitary condition, has been successfully met by spreading ashes over the surface of these runs to a depth of about three inches.

It is found to be necessary, from time to time, to add a fresh, thin coating of ashes over the surface of these runs, as the chicks make way with such a large quantity.

Eggs for Hatching

As is stated in another chapter, the layers are disposed of immediately at the close of their first laying season, or when the birds are between fifteen and sixteen months old. This is true of all the stock that is kept for the production of eggs for domestic purposes; but each season a sufficient number of the best pullets are selected and transferred to the breeding house, for the eggs of yearling hens hatch stronger and better chicks than those of pullets.

No trap nests are used on this farm for the reason that it is believed they interfere to a greater or lesser extent with the laying of the birds. The Leghorn is a very nervous fowl, and the closing of the

---

Professor Gowell preached: "Keep clean water, charcoal, granulated bone, oyster shell and sharp grit always before the chicks; and cracked bone, oyster shell, grit and water before the hens all the time."—EDITOR.

confining door of the trap nest always has a tendency to keep her in a nervous state. This is not conducive to a big egg yield.

Notwithstanding this, the proprietors of the farm believe that they are able with close accuracy to select their best layers for the breeding pen. Indeed, when the egg yield given in detail in a previous chapter is considered, it would seem that there were no drones in the laying houses. Every bird on the farm carries a year numbered leg band.

With these yearling hens are mated carefully-selected cockerels, one for every fifteen hens. No attempt is made to divide these breeders into flocks, but all run together in the one room and yard. Cockerels are used rather than cocks because experience has proven that they throw a larger percentage of pullets. The experience on Sunny Slope Farm backs up this theory, only one-third of the chicks hatched in the last season being males. It is also believed that cockerels produce a larger percentage of fertility than can be obtained from older males.

The birds are not mated until within ten days or two weeks of the time that it is desired to start the incubators. Experience has proven that Leghorn eggs are fertile within three to five days after mating.

The eggs produced by birds mated this way are gathered at regular intervals and placed in turning machines, being carried so as to lie on the end and not on side. They are turned regularly every day until they are placed in the incubator. This is done to prevent the germs adhering to the side of the shell.

The sooner an egg goes into the incubator after being laid the better. At Sunny Slope Farm an effort is made not to have the eggs over two days old.

Some breeders believe that a larger percentage of fertility may be obtained if each male was given his own mates. It would involve considerably more labor, however, and as the eggs from the breeder pens on this farm have given better than a 90 per cent. fertility, and some have gone as high as 95 per cent. With strong, lively, fluffy chicks, it does not seem that any mistake is being made in placing all the breeders in one room.

Last season the first chicks were brought out in March, but many of these molted after they had laid for a few weeks in the fall. It has been decided that in the future all the stock on this farm will be hatched between the 10th of April and the 10th of June.

This is not only the best period to hatch Leghorns for winter eggs, but it is the natural period, and the eggs are always more hatchable and produce better chicks—chicks that live and thrive—than either earlier or later.

Cleanliness

Only five birds were lost from all causes after the pullets were removed from the colony to the laying houses on Sunny Slope Farm last season. This is attributed to the absolute cleanliness which is maintained there. Cleanliness is a vital element in chicken raising.

Sunny Slope Farm does not use trap nests, believing that it makes the Leghorns too scary. The experience of the writer, however, not only proves that Leghorns very quickly get over their nervousness when trapped and handled a few times, but become very tame.

Professor Gowell was a staunch believer in the trap nest, and had between 400 and 500 nests for the 2,000 to 2,500 hens at the Maine Station Farm, and 400 traps on his own Go-well Farm. It is the only absolutely sure way to pick out drones.

Sunny Slope Farm estimates that to trap nest 5,000 hens would cost at least $1,000 a year labor, which is one reason why they do not do it.—EDITOR.

All who visit Sunny Slope Farm mark the order of cleanliness that is kept. The writer never saw cleaner houses, and this is remarkable, too, considering the large number of fowls that are housed all the time. Without strict cleanliness it would be impossible to keep such a number of birds in the pink of condition.—EDITOR.

The dropping boards in the laying and breeding houses are cleaned every day, the droppings being carefully stored in a shed specially provided for the purpose. The drinking fountains in these houses are washed and scoured with a brush every morning. This removes all the slime which naturally clings to the sides from the water. The nests are gone over every day, and any filth which may have been taken into them by the fowl is removed. Excelsior has been found capital material to use in the nest boxes. It is clean and sanitary.

The birds are not allowed to roost anywhere except on the perches provided for that purpose. This prevents the birds inclined to steal their roosts from befouling any section of the house except the dropping boards, and helps to maintain the general cleanliness.

Every few days the canvas drops which act as windows are brushed with a stiff whisk-broom to remove any dust adhering to them and which may prevent the free access of the outside air.

No disinfectants or lice killers are used, for the reason that they have never been required. The absolute dryness of the house probably makes it uncomfortable for lice. At any rate, they have never yet appeared in the houses on this farm.

When the laying stock is disposed of in the fall the laying houses are thoroughly cleansed. All the litter is removed and the floors are swept. Then the entire interior is gone over with a mixture of kerosene and crude carbolic acid. New litter is placed on the floor and the houses are ready for another flock of laying pullets. At the same time the nest boxes are all removed, cleansed and painted with the above-named mixture.

The colony houses are all cleaned out at least twice a week. There are no roosts in these houses, and consequently the litter has to be removed at every cleaning. The hover parts of the runs in the brooder house are scraped and thoroughly cleaned every day while they are in use. In the yards covered with litter the floor is swept and everything removed every three weeks, or at the time the chicks are moved to the cold hovers to make room for another lot. The alleyway in this house is regularly swept and the hot-water pipes are frequently dusted.

At the end of each hatch the incubators are thoroughly gone over. The lower diaphragms, drawers and trays are carried outdoors and laid in the sun. When thoroughly dry they are swept with a stiff brush until every foreign substance is removed from them.

All this detail is gone over regularly. Older poultrymen do not think such close attention to this matter of cleanliness is necessary, but here there are very fixed opinions thereon.

Punctuality and Regularity

It has been said that the hen is a systematic animal. One thing is very certain—she works on time. Close attention to this characteristic of the hen has been one of the important factors in bringing success to Sunny Slope Farm. Everything there is done by the clock—a large eight-day one, which hangs in a prominent place in the workshop.

Following is the day's schedule:

Between five-thirty and six o'clock every morning in summer, and as soon as it is light in winter, the attendants open the house and put water—boiling hot in winter—into the drinking fountains.

At nine o'clock green food is fed, and the first gathering of eggs follows.

This is another great secret of success. It was no easy task to get such a large farm like Sunny Slope down to such a good system. The poultrymen who will "go and do likewise" will find that in poultry culture there are no more important acts than punctuality and regularity.—EDITOR.

At ten-thirty o'clock green food is given the cockerels.

At eleven-thirty oats are fed, sometimes mixed with buckwheat, and the second gathering of eggs follows.

At two-thirty o'clock the third gathering of eggs is made. This is always the principal collection of the day.

At three o'clock in the winter months mash is placed in the troughs. In the summertime it is fed at four o'clock.

At three-thirty o'clock the cockerels are given their mash ration.

At five o'clock in the summer grain is fed in the litter. In the wintertime this is varied to make the feeding one hour before sunset, in order to give the fowls plenty of opportunity to fill up before it becomes dusk.

According to the weather the houses are closed for the night, and at dusk a final careful search for eggs is made, not only in the nest boxes, but particularly in the litter.

At seven-thirty o'clock the houses are again visited and all birds not so roosting are placed on the perches.

This schedule is adhered to rigidly throughout the year, nothing whatever being allowed to interfere with it.

The Corning Method Applies to Small Plants as Well as Large

The question is often asked whether the Sunny Slope Farm methods can be successfully used in handling a few hens, as well as with a large number. The start at Sunny Slope Farm was made with a breeding pen of thirty Single Comb White Leghorn yearling hens and three cockerels—one 390 egg capacity incubator, and three portable brooders.

While it is believed that a width of 16 feet is the standard for poultry houses in most climates the length of the house is a matter of individual requirement. It must, however, be borne in mind that in a house 16 feet wide by, say, 20 feet long, one cannot carry as many birds, in proportion to the floor space, as in a house, say, 60 feet long. And the same is true as between a house 60 feet long and one 100 feet long. This because, in a long house, while the square feet of floor space, per hen, is small, this is offset by the fact that each hen has the range of the entire floor of the house.

This general system of feeding can be carried on for a small flock of hens, by substituting beef scraps for cut bone—where there is no bone cutter—and mixing the mash with paddles, in a tub or other convenient vessel. Small bone cutters, operated by hand, cost but little.

A man living near Sunny Slope Farm, who carries about 100 pullets, always had a very small egg yield, until two winters back, when he adopted Sunny Slope Farm methods, beginning in January. Even with this midwinter start, his egg production increased in what, to him, was a most astonishing degree. The following autumn he bought a small pen (15 hens and one cock) of Sunny Slope Farm stock, and bred from them the following season. He increased his flock of young chicks by obtaining hatching eggs from the same source, and thus produced pullets whose parents and grandparents had been bred under the Corning System. His present egg yield largely exceeds all previous records.

To further illustrate the working of this system on a still smaller scale: A gentleman living in Bound Brook had a mixed pen (hens and pullets) of 18 Barred Rocks and R. I. Reds.

The pullets were hatched early and should have been laying by October. The hens molted early and looked well, but neither hens nor

INTERIOR OF LAYING HOUSE NO. 3

Fifteen hundred pullets at work

pullets laid any eggs. The owner having "tried out" various "sure egg producing" feeds and methods, tried the Sunny Slope Farm way, the latter part of December, 1908. He reported that after ten days' hard work in the litter the birds began to lay and continued laying through the winter and summer.

The above illustrations—taken from many others—seem to fully answer the question: How will this system work with a small flock?

Index

A
	PAGE
Advanced prices	12
Air, fresh	22
Air-space in walls	33
Alfalfa	53
Animal food	51
Ashes	48, 55, 56
Atmosphere, dry	43
Automatic drinking fountains	55
Awning to exclude sun	29

B
Band, leg	57
Beef scrap	45, 47, 48, 49
Bins, food	39
Boiling-hot water	58
Bone cutter	39, 59
Bone	12, 47, 49, 52
Bowel looseness	45
Bowel regulator	56
Boxes, paper	14
Bran, wheat	12, 48, 49
Breeders	13
Breeding house	37, 38
Breeding pens	19, 56
Breeding stock, feeding	51
Breed to keep	18
Broilers	13, 49
Brooder house	20, 23, 25, 27, 44
Brown eggs	18
Buckwheat	48

C
Carbolic acid, crude	58
Cartons, paper	13
Ceiling boards	33
Ceiling height	31, 41
Cellar, incubator	23, 25, 41
Cement floors	23, 25
Cement walls	23
Charcoal	12, 45, 47, 55, 56
Cheese cloth	33
Chicks, care of	44, 45
Chimneys, ventilating flues in	23
Circulation of air	23
Cleanliness	56, 57, 58
Close partition house	22
Cloth curtain	33
Clover, cut	12, 48, 53
Clover cutter	47
Coal ashes	48, 55, 56
Cockerel house	37, 38
Cockerels	20
Cockerels, feeding	49
Cold air	28
Cold storage eggs	14
Colds	22, 23
Cold water	54, 55
Collecting eggs	17
Colony houses	21, 29, 31
Concrete floor	43
Concrete sills	25
Condiments	52
Corn	12, 19, 45, 47
Corn meal	48, 49

D
	PAGE
Corning method applied to small plants	59
Cost of feeding	12, 13
Cost of housing poultry	20, 21
Cotton-duck	23, 29, 34
Crops, examining	45
Crows	44
Curtain-front house	22

D
Daily shipments	15
Dampness	22, 23
Dark room	25
Death, premature	22
Diarrhea	53
Digestion, impaired	45
Disease in large flocks	20
Disinfectants	58
Doors	27, 29, 36
Dopy hens	52
Draughts	22, 23, 27, 28, 33, 35
Drinking water	54, 55
Drones, to pick out	57
Dropping boards	36, 37, 58
Droppings	27, 39, 58
Dry grains	45
Dry meal mixture	47
Dust	37, 58
Dust bath	37
Dusting	58

E
Early Pullets	19
Economy in building	22
Economy of space, etc.	20, 21
Egg farming	9
Egg record	17
Egg three-fourths water	55
Eggs, cold storage	14
Eggs, fertile	25, 41
Eggs, flavor of	53
Eggs, fresh	13
Eggs for hatching	55, 56
Eggs, gathering	58, 59
Eggs, infertile	25, 41
Eggs, packing	12
Eggs, price of	13
Eggs, scaled	17
Eggs, sterile	19
Eggs, testing	25, 41
Eggs, turning	39, 43
Eggs, weight of	15
Eggs, winter	43, 57
Egg shells	17
Egg shells, color of	18
Emergency pipes	25
Engine, gasoline	39
Entrances	36
Excelsior for nests	58
Exercise	22
Exposure	38

F
False floor	27
Fat, over	19
Fattening pens	21, 49

	PAGE
Feed hoppers	56
Feed house	39
Feeding	59
Feeding breeding stock	51
Feeding chicks	45, 46
Feeding cockerels	49
Feeding cockerels for broilers	49
Feeding, cost of	12
Feeding during molting	51
Feeding laying pullets	48
Feeding on range	47
Feeding pullets	47, 48
Feeding troughs	47
Fences	10, 37
Fertile eggs	25
Fertility	56, 57
Flavor of eggs	53
Flocks, size of	20, 21
Floor, concrete	43
Floor joists	33
Floor space	21, 29, 31, 59
Floors	33
Flues, ventilating	23
Food bins	39
Food, green	53, 55
Foul air	28
Foundation	31
Fountains, automatic	55
Fountains, drinking	47, 48
Freezing	25
Fresh air	22, 23, 28
Fresh cut bone	52
Fresh eggs	13
Full crop	52

G

Gases, poisonous	43
Gasoline	12
Gasoline engine	39
Gate	27
Gathering eggs	17, 58, 59
Germ adhering to shell	57
Gizzard needs work	45
Glass	33
Gluten meal	48, 49
Go-well Farm	6, 7
Gowell, Gilbert, M.	6, 9
Gowell mash	48
Granulated bone	56
Grass and clover	53
Grassy range	47
Green bone	39
Green bone, cut	40
Green food	12, 51, 53, 55
Grit	12, 45, 47, 48, 55, 56
Ground oats	12, 49

H

Half breeds	18
Handling chicks	44
Hatching	41, 43
Hatching, eggs for	56
Hawks	44
Hens, feeding molting	51
Hot water	54, 55
Hot-water heater	25
Hot-water pipes	28, 29
House room	20
Housing poultry, cost of	20
Hovers	20, 25, 27, 44

I

Ice house	39
Inbreeding	18, 19
Incubator cellar	23, 25, 41
Incubators	39

	PAGE
Infertile eggs	25
Insects	47, 48

K

| Kerosene | 58 |

L

Labor	12
Lamps	27, 39, 43
Large flocks, economy of	20
Laying houses	21, 22, 31
Laying pullets, feeding	48
Laying too soon	19
Leg band	57
Level, spirit	39
Lice killers	58
Lime	48, 56
Limestone	56
Line breeding	19
Linseed meal	48
Litter	36, 37, 47, 48, 58

M

Machinery	39
Manure	13
Markets	13
Marketing the eggs	17
Market, off to	16
Market quotations	14
Marking chicks	19
Mash	39, 45, 48, 49, 51, 52
Mash, Gowell	48
Meal	12
Meat chopper	45
Meat meal	49
Middlings	48, 49
Milhen eggs	14
Milk, skim	49
Mixed pen	59
Mixing machine	39
Moisture	3, 41, 43
Molting	19, 44, 51
Mongrels	18
Months, best hatching	43
Morning or night mash	52
Morning ration	48
Musty wheat	45

N

Narrow houses	22
Nests	37, 38
Nests, trap	56, 57
Netting, wire	31, 36
New York quotations	14
Nursery	20, 25, 28

O

Oats	12, 45, 48, 54, 55
Oats, green for food	53
Oats, sprouted	55
Oil meal	49
Operating incubators	39
Overeating	45
Overfat hens	51
Oxygen	22, 23
Oyster shells	12, 47, 55, 56

P

Packing eggs	12
Paddle for mixing	49
Paper boxes	8, 13, 17
Paper, building	33, 36
Partitions	36

	Page
Pens	27, 49
Percentage of loss	10, 44
Perches	36
Philadelphia prices	14
Pipes, hot-water	28, 29
Plan of Sunny Slope Farm	10, 11
Porridge	49
Portable brooders	59
Portable houses	47
Postage	13
Posts under houses	22
Premium prices	13
Processed oats	55
Pullets, feeding	47, 48
Punctuality	58
Pure air	29
Pure water	55

Q

Quality	17

R

Rafters	33
Range	19, 29, 47
Ration, morning	48
Ratproof building	25, 33
Rats	44
Record, keeping a	41
Regularity	58
Retail trade	12
Rolled oats	45
Roofing	25, 29, 31, 33, 34, 47
Roofing paper	25
Roosts	36, 37
Roup	23
Runs, keeping clean	56
Runway	27

S

Salt meat	53
Scraps, beef	45, 47, 48
Scratching in litter	47
Screenings	45
Sealing egg packages	17
Securing a market	15
Sex, age to determine	20
Shade	10, 29, 48
Shell	12, 18
Shells, feeding	45
Shells, need lime	56
Sills	25, 33
Single-room laying house	22
Site, selecting a	23
Six dollars per hen	12
Size of eggs	18
Size of flocks	20, 21
Skim milk	49, 51
Skunks	33
Sliding door	27
Slime in drinking fountains	58
Sluggish hens	52
Small egg yield, increasing	59
Small plants	59

	Page
Snow storms	37
Sour wheat	45
Space, floor	21, 59
Spirit level	39
Sprouted oats	54, 55
Stamina, inherited	22
Sterile eggs	19
Stormy weather	37
Strain, importance of	18
Straw	47, 48
Sunlight	22, 23
Sure egg-producers	61
Sweeping	58

T

Tables	39, 41
Tag, record	44
Tainted meat or bones	52, 53
Tar paper	36
Temperature	23, 25, 29, 41, 43
Testing eggs	25, 41
Time to feed	49
Trade, how attracted	12, 17
Trap nests	55, 56, 57
Trays, incubator	39
Troughs	47, 49
Turning eggs	39, 43, 57
Two hundred eggs per year	7

V

Ventilating windows	25, 28
Ventilation	22, 23, 28, 30, 41
Vestibule entrance	25
Vigor, constitutional	22
Vitality	22

W

Water, drinking	47, 54, 55
Water shed	33
Weaklings	41
Weasels	33
Web, punching	19
Weight of eggs	15
Wheat bran	48, 49
Wheat chaff	45
Wheat, sour or musty	45
Wheat straw	48
When to hatch	43
White eggs	18
Whole wheat	48
Wide houses	22
Windmill	10
Windows	27, 28, 29, 33, 41
Windows, ventilating	25, 28
Wind storms	37
Winter eggs	43, 57
Winter wheat for green food	53
Wire netting	31, 36
Workshop	39
Worms	47

Y

Yearling hens	19

www.ingramcontent.com/pod-product-compliance
Lightning Source LLC
Chambersburg PA
CBHW062335220526
45469CB00008B/2725